高等学校水利类统编教材

# 水利专业英语

# ENGLISH FOR WATER RESOURCES


主　编　迟道才　周振民

副主编　杨路华　马永胜　孙西欢　徐　丹

中国水利水电出版社

www.waterpub.com.cn

## 内 容 提 要

本书以培养学生专业英语阅读与翻译能力为主要目标，内容包括水文学、水资源、水环境、灌溉排水、水工建筑物、水力发电、水利机械、工程施工、工程材料、水利经济、水利规划、工程测量、工程招标、科技写作等方面的文章。全书由32篇课文、32篇阅读材料和1个讲座组成。全部课文均附有参考译文。

本书可以作为水利水电工程、农业水利工程、农业水土工程、水文与水资源等专业本科生和研究生的专业英语教材或课外阅读材料，也可以供从事相关专业的工程技术人员、管理人员和教师自学使用。

**图书在版编目（CIP）数据**

水利专业英语＝English for Water Resources：英文/迟道才，周振民主编．—北京：中国水利水电出版社，2006（2022.6重印）
高等学校水利类统编教材
ISBN 978-7-5084-3501-5

Ⅰ．水… Ⅱ．①迟…②周… Ⅲ．水利工程-英语-高等学校-教材-英文 Ⅳ．H31

中国版本图书馆CIP数据核字（2005）第155598号

| 书 名 | 高等学校水利类统编教材<br>**水利专业英语（ENGLISH FOR WATER RESOURCES）** |
|---|---|
| 作 者 | 主编 迟道才 周振民 |
| 出版发行 | 中国水利水电出版社<br>（北京市海淀区玉渊潭南路1号D座 100038）<br>网址：www.waterpub.com.cn<br>E-mail：sales@mwr.gov.cn<br>电话：(010) 68545888（营销中心） |
| 经 售 | 北京科水图书销售有限公司<br>电话：(010) 68545874、63202643<br>全国各地新华书店和相关出版物销售网点 |
| 排 版 | 北京文瑞达数据技术有限公司 |
| 印 刷 | 清淞永业（天津）印刷有限公司 |
| 规 格 | 184mm×260mm 16开本 14.25印张 356千字 |
| 版 次 | 2006年1月第1版 2022年6月第9次印刷 |
| 印 数 | 20001—22500册 |
| 定 价 | **45.00**元 |

# 前　　言

国家教育部颁布的“大学英语教学大纲”把专业英语阅读列为必修课而纳入英语教学计划，强调通过4年不间断的英语教学使学生达到顺利阅读专业刊物的目的。根据这个精神，按照教育部新的学科和专业调整目录，编写了《水利专业英语》教材，以满足高等院校水利水电工程、农业水利工程、水文与水资源以及其他相关专业的专业英语教学需要和从事相关专业的工程技术人员和管理人员学习专业英语的要求。

本书涉及的内容包括水文学、水资源、水环境、灌溉排水、水工建筑物、水力发电、水利机械、工程施工、工程材料、水利经济、水利规划、工程测量、工程招标、科技写作等方面。考虑到英文摘要写作的重要性和特殊性，本书将中文论文中的英文摘要的写作要求与写作方法作为一个专题进行讲解，意在提高科技论文英文摘要的翻译水平，这也是本书的特色之一。

全书由32篇课文、32篇阅读材料和1个讲座组成，全部课文均附有参考译文。本书按照水利工程一级学科进行选材，除参考文献所列书目外，还参考了其他兄弟院校编写的专业英语内部教材。本书语言规范，题材广泛，内容由浅入深，适应性和针对性强，便于自学，符合素质教育的基本要求。

参加本书编写工作的有沈阳农业大学迟道才、王瑄、李智，华北水利水电学院周振民，河北农业大学杨路华，东北农业大学马永胜，太原理工大学孙西欢、马娟娟，辽宁师范大学外国语学院徐丹。迟道才、周振民担任主编，杨路华、马永胜、孙西欢、徐丹担任副主编，全书由迟道才负责统稿。

由于编者水平所限，书中难免存在不足和错误，恳请广大读者批评指正。

编者

2005年11月

# 目　　录

# Lesson 1 Importance of Water

Water is the best known and most abundant of all chemical compounds occurring in relatively pure form on the earth's surface. Oxygen, the most abundant chemical element, is present in combination with hydrogen to the extent of 89 percent in water. Water covers about three fourths of the earth's surface and permeates cracks of much solid land. The polar regions are overlaid with vast quantities of ice, and the atmosphere of the earth carries water vapor in quantities from 0.1 percent to 2 percent by weight. It has been estimated that the amount of water in the atmosphere above a square mile of land on a mild summer day is of the order of 50,000 tons.

All life on earth depends upon water, the principal ingredient of living cells. The use of water by man, plants, and animals is universal. Without it there can be no life. Every living thing requires water. Man can go nearly two months without food, but can live only three or four days without water.

In our homes, whether in the city or in the country, water is essential for cleanliness and health. The average American family uses from 65,000 to 75,000 gallons of water per year for various household purposes.

Water can be considered as the principal raw material and the lowest cost raw material from which most of our farm produce is made. It is essential for the growth of crops and animals and is a very important factor in the production of milk and eggs. Animals and poultry, if constantly supplied with running water, will produce more meat, more milk, and more eggs per pound of food and per hour of labor. For example, apples are 87% water. The trees on which they grow must have watered many times the weight of the fruit. Potatoes are 75% water. To grow an acre of potatoes tons of water is required. Fish are 80% water. They not only consume water but also must have large volumes of water in which to live. Milk is 88% water. To produce one quart of milk a cow requires from 3.5 to 5.5 quarts of water. Beef is 77% water. To produce a pound of beef an animal must drink many times that much water. If there is a shortage of water, there will be a decline in farm production, just as a shortage of steel will cause a decrease in the production of automobiles.

In addition to the direct use of water in our homes and on the farm, there are many indirect ways in which water affects our lives. In manufacturing, generation of electric power, transportation, recreation, and in many other ways, water plays a very important role.

Our use of water is increasing rapidly with our growing population. Already there are acute shortages of both surface and underground waters in many localities. Careless pollution and contamination of our streams, lakes, and underground sources has greatly impaired the quality of the water which we do have available. It is therefore of utmost importance for our future that good conservation and sanitary measures be practiced by everyone.

## New Words and Expressions

1.permeate *v.* 渗入，渗透
2.beef *n.* 牛肉
3.polar *a.* 地极的，（南、北）极的
4.decline *n.* 下降，下落
5.overlay *v.* 覆盖 overlaid, overlaid
6.manufacturing *n.* 制造，制造业
7.mild *a.* 温暖的
8.acute *a.* 尖锐的，剧烈的
9.ingredient *n.*（混合物的）成份，要素
10.locality *n.* 地点，地区
11.cleaniness *n.* 清洁
12.contamination *n.* 污染，弄脏
13.household *n.* 家庭
*a.* 家用的
14.impair *vt.* 损害；障碍
15.raw *a.* 原状的，原始的
16.available *a.* 可利用的
17.poultry *n.* 家禽
18.conservation *n.* 保护
19.acre *n.* 英亩
20.sanitary *a.* 卫生的，保健的
21.consume *v.* 消耗，消费
22.to the extent of （达）到…的程度
23.quart *n.* 夸（脱）[容量单位，1 夸（脱）=1/4 加仑=1.14 升]
24. (be) of the order of 大约，约为
25.(be) essential to (for) 对…（来说）是必需地（必不可少的）
26.compound *n.* 化合物

## Reading Material Water Resources of the Yellow River Basin

Water resources of the Yellow River basin include surface water resources and ground water resources. The replenishment water source is precipitation.

### 1 Surface Water Resources

The Yellow River basin climate is continental monsoon type. In winter, it is controlled by polar cold air mass of more northwest wind and less rain and snow. In summer, it is effected by sub-tropical high pressure of the west Pacific with abundant rainfall. The mean annual precipitation of the whole basin is 466 mm. The average natural annual runoff of the river is 58 billion $m^3$, which is less than that of the Yangtze River, Pearl River and Songhua River, making up only 2% of the country and ranking the fourth of the seven large rivers in China. The annual average runoff per capita is $573m^3/s$ and 324 $m^3$ per mu of cultivated land, making up 25% and 17% average per capita and per mu of the country. The Yellow River is deficient in water resources.

The distribution of annual runoff of the river is uneven and 56% of water comes from the upper most section above Lanzhou. The annual runoff of the river changes greatly with seasons and 60% of that concentrates in the flood season from July to October. The annual runoff between years also varies greatly and the ratio of the maximum annual runoff to the minimum annual runoff at Huayuankou Station is 3.4. The Yellow River basin also has the characteristics of continuous long period of dry years, such as during the period from 1922 to 1932 that the river was dry for eleven years continuously and the annual runoff was 30% less than that of normal years. This brings a lot of unfavorable effects to the utilization of water resources. In addition, the high sediment concentration of water also has increased the difficulty of the development and utilization of water resources.

## 2 Ground Water Resources

The ground water resources in the shallow ground water with an average degree of mineralization is less than 3 g/l. The net volume of ground water resources which is not duplicated with the surface water resources of the basin is 13.9 billion $m^3$. Of which, about 70% of net exploitation is concentrated in Hetao Plain in Ningxia—Inner Mongolia Autonomous Region, in the Fen-Wei Basin in the middle basin and in the plain adjacent to the lower Yellow River.

## 3 Total Volume of Water Resources

The total volume of water resources is the sum of surface runoff and ground water resources. The total volume of water resources of the Yellow River is 71.9 billion $m^3$.

## 4 Water Quality

The natural water quality of the main river is good. But, with the development of agriculture and industry, and with the increase of city population, the pollution of the river has become serious day-by-day. At present, the total volume of waste and polluted water discharged into the Yellow River has increased from 2.17 billion tons in the beginning of 1980's to 4.2 billion tons in 1990's. The large pollution sources along the river are more than 300 and the pollution degree of water quality of the river has ranked the second among the seven large rivers in China. The more serious water quality pollution of the main river is mainly distributed in the reaches of Lanzhou, Baotou and Tongguan. For tributaries, that is mainly distributed in the sections of Xining on the Huangshui River, Taiyuan on the Fenhe River, Baoji and Xi'an on the Weihe River, Luoyang on the Luohe River and Laiwu on the Wenhe river.

# New Words and Expressions

1.basin *n.* 盆[洼，泊]地，流域

2.replenishment *n.* 再倒[装]满，（再）补给[充]，充实[满，填]，供给

3.monsoon *n.* 季（节）风，贸易风

4.tropical *a.* 热带[地方，似]的，回归线下的，酷[炎]热的，热烈的

5.the Yangtze River 长江

6.the Pearl River 珠江

7.capita caput 的复数

per capita 每人（口）

8.sediment *n.* 沉积物，沉淀物，泥沙

9.mineralization *n.* 矿化[成矿]（作用）

10.duplicate *vt.* 重叠，双折，加倍，重复

11.ranking *n.* 顺序，序列，排列，等级

# Lesson 2 The Hydrologic Cycle

In nature, water is constantly changing from one state to another . The heat of the sun evaporates water from land and water surfaces. This water vapor (a gas), being lighter than air, rises until it reaches the cold upper air where it condenses into clouds. Clouds drift around according to the direction of the wind until they strike a colder atmosphere. At this point the water further condenses and falls to the earth as rain, sleet, or snow, thus completing the hydrologic cycle.

The complete hydrologic cycle, however, is much more complex. The atmosphere gains water vapor by evaporation not only from the oceans but also from lakes, rivers, and other water bodies, and from moist ground surfaces. Water vapor is also gained by sublimation from snowfields and by transpiration from vegetation and trees.

Water precipitation may follow various routes. Much of the precipitation from the atmosphere falls directly on the oceans. Of the water that does fall over land areas, some is caught by vegetation or evaporates before reaching the ground, some is locked up in snowfields or ice-fields for periods ranging from a season to many thousands of years, and some is retarded by storage in reservoirs, in the ground, in chemical compounds, and in vegetation and animal life.

The water that falls on land areas may return immediately to the sea as runoff in streams and rivers or when snow melts in warmer seasons. When the water does not run off immediately it percolates into the soil. Some of this groundwater is taken up by the roots of vegetation and some of it flows through the subsoil into rivers, lakes, and oceans.

Because water is absolutely necessary for sustaining life and is of great importance in industry, men have tried in many ways to control the hydrologic cycle to their own advantage. An obvious example is the storage of water behind dams in reservoirs, in climates where there are excesses and deficits of precipitation (with respect to water needs) at different times in the year. Another method is the attempt to increase or decrease natural precipitation by injecting particles of dry ice or silver iodide into clouds. This kind of weather modification has had limited success thus far, but many meteorologists believe that a significant control of precipitation can be achieved in the future.

Other attempts to influence the hydrologic cycle include the contour plowing of sloping farmlands to slow down runoff and permit more water to percolate into the ground, the construction of dikes to prevent floods and so on. The reuse of water before it returns to the sea is another common practice. Various water supply systems that obtain their water from rivers may recycle it several times (with purification) before it finally reaches the rivers mouth.

Men also attempt to predict the effects of events in the course of the hydrologic cycle. Thus, the meteorologist forecasts the amount and intensity of precipitation in a watershed, and the hydrologist forecasts the volume of runoff.

## New Words and Expressions

1.hydrologic *a.* 水文的，水文学的
  hydrologic cycle 水循环
2.condense *v.* 凝结
3.drift *v.* 飘动，飘浮
4.sleet *n.* 雨夹雪，冻雨
5.moist *a.* 潮湿的
6.sublimation *n.* 升华，升华作用
7.transpiration *n.* 蒸腾，蒸腾作用
8.vegetation *n.* 植物
9.precipitation *n.* 降水，降雨
10.route *n.* 路程，道路
11.retard *vt.* 停滞
12.storage *n.* 贮藏
13.runoff *n.* 径流
14.percolate *v.* 渗透
15.root *n.* 根
16．subsoil *n.* 下层土，底土
17.sustain *vt.* 维持，使…生存下去
18.excess *n.* 过量，过剩
19.deficit *n.* 欠缺，不足
20.inject *vt.* 注射，喷射
21.iodide *n.* 碘化物
  silver iodide 碘化银
22.modification *n.* 改善，改变
23.meteorologist *n.* 气象学家
24.contour *a.* 沿等高线修筑的
25.plowing *n.* 耕地
26.sloping *a.* 倾斜的
27.dike *n.* 堤，坝
28.purification *n.* 净化，澄清
29.predict *vt.* 预测
30.intensity *n.* 强度
31.watershed *n.* 流域
32.hydrologist *n.* 水文学家
33.lock up 封闭，潜藏
34.take up 吸收，溶解
35.to one's advantage (to the advantage of) 对…有利
36.with respect to 根据，关于，就…而论
37.thus far 迄今

## Reading Material The Hydrologic Cycle

The hydrologic cycle is the process whereby water is converted from its liquid or solid state into its vapor state. As a vapor the water is capable of traveling considerable distances from its source prior to recondensing and returning to earth as precipitation. Thus the hydrologic cycle is a complex, interrelated system involving the movement of atmospheric surface (marine and fresh), and groundwater throughout various regions of the world. It is the hydrologic cycle that is solely responsible for the world's precipitation. And it is this precipitation, falling on the terrestrial and surface freshwater environments, that is the sole source of the earth's supply of fresh water.

The hydrologic cycle may consist of either a long or various short cycles. In the short cycles, water may evaporate from either marine or freshwater systems, condense almost immediately, and return as precipitation to the same system. Another variation of a short cycle is the precipitation and subsequent evaporation of water from land surfaces, followed by its condensation and return as precipitation to the land, followed by reevaporation, and so on.

In the long cycle the major source of water vapor is the world's oceans, which contain 97.3% of the earth's waters. In this cycle a portion of the water evaporates and forms clouds that move inland. The water vapor then cools and returns to earth as precipitation. It is estimated that only 0.007% of the oceanic water is distributed to terrestrial areas annually. This water will ultimately return to the oceans through river and groundwater flow. Since precipitation may occur close to the source of initial evaporation or thousands of miles away, the water may remain in the vapor state for variable times (a few hours to a few weeks). The average residence time for water to remain in the atmosphere is considered to be 10 days.

It is as well to have a clear idea of the scale of the events that are being discussed. Table2.1 lists estimates of the amounts of water involved in the hydrological cycle and the proportion (in percentages) of the total water on earth involved in each part of it.

Table 2.1 Estimated earth's water inventory

<table>
<tr><th>Location</th><th>Volume<br>($\times 10^{12}m^3$)</th><th>Percentage (out of total water)<br>(%)</th></tr>
<tr><td>Fresh-water lakes</td><td>125</td><td rowspan="4">0.62</td></tr>
<tr><td>Rivers</td><td>1.25</td></tr>
<tr><td>Soil moisture</td><td>65</td></tr>
<tr><td>Groundwater</td><td>8,250</td></tr>
<tr><td>Saline lakes and inland seas</td><td>105</td><td>0.008</td></tr>
<tr><td>Atmosphere</td><td>13</td><td>0.001</td></tr>
<tr><td>Polar ice-caps, glaciers and snow</td><td>29,200</td><td>2.1</td></tr>
<tr><td>Seas and oceans</td><td>1,320,000</td><td>97.25</td></tr>
<tr><td>Total</td><td>1,360,000</td><td>100.0</td></tr>
</table>

Of the 0.6 percent of total water that is available as fresh water, about half is below a depth of 800m and so is not practically available on the surface. This means that the stock of the earth's fresh water that is obtainable one way or another for man's use is about $4\times10^6km^3$ and is mainly in the ground. Spread over the earth's land surface it would be about 30m deep.

The four processes with which the hydrologist is mainly concerned are precipitation, evaporation and transpiration, surface runoff or stream flow, and groundwater flow. He needs to be able to interpret data about these processes and to predict from his studies the most likely quantities involved in the extreme cases of flood and drought. He must be able also to express an opinion about the likely frequency with which such events will occur, since it is on the frequency of certain values of extreme events that much hydraulic engineering design is based.

## New Words and Expressions

1.recondense *v.* 再凝聚

2.interrelate *v.* 相互有关

3.solely *ad.* 独自，单独

4.terrestrial *a.* 地球上的，陆地的

5.variation *n.* 变化，改变
6.subsequent *a.* 后来的，接着发生的
7.reevaporation *n.* 再次蒸发
8.inland *ad.* 向内地 *a.* 内地的，内陆的
9.oceanic *a.* 海洋的，大洋的
10.distribute *vt.* 分布，散布
11.annually *ad.* 每年
12.initial *a.* 最初的，初期的
13.variable *a.* 不同的，不定的
14.residence *n.* 滞留，停留
15.inventory *n.* 总量，详细目录
16.stock *n.* 贮存，贮藏
17.interpret *v.* 解释，阐明
18.datum *n.* 资料，数据
19.drought *n.* 干旱
20.hydraulic *a.* 水力的，水工的
hydraulic engineering 水利工程
21.prior to 在…以前，优先于
22.be responsible for 是（造成）…的主要原因，导致，引起
23.followed by 后面有，后面是，再加上
24.and so 因此，所以
25.spread over 散布，分布

# Lesson 3  Hydrology

## 1  History

The first hydraulic project has been lost in the mists of prehistory. Perhaps some prehistoric man found that pile of rocks across a stream would raise the water level sufficiently to overflow the land that was the source of his wild food plants and thus water them during a drought. Whatever the early history of hydraulics, abundant evidence exists to show that the builders understood little hydrology. Early Greek and Roman writings indicated that these people could accept the oceans as the ultimate source of all water but could not visualize precipitation equaling or exceeding stream-flow. Typical of the ideas of the time was a view that seawater moved underground to the base of the mountains. There a natural still desalted the water, and the vapor rose through conduits to the mountain tops, where it condensed and escaped at the source springs of the streams. Marcus Vitruvius Pollio (ca. 100 B. C.) seems to have been one of the first to recognize the role of precipitation as we accept it today.

Leonardo da Vinci (1452-1519) was the next to suggest a modern view of the hydrologic cycle, but it remained for Pierre Perrault (1608-1680) to compare measured rainfall with the estimated flow of the Seine River to show that the stream-flow was about one-sixth of the precipitation. The English astronomer Halley (1656-1742) measured evaporation from a small pan and estimated evaporation from the Mediterranean Sea from these data. As late as 1921, however, some people still questioned the concept of the hydrologic cycle.

Precipitation was measured in India as early as the fourth century B. C., but satisfactory methods for measuring stream-flow were a much later development. Frontinus, water commissioner of Rome in A. D. 97, based estimates of flow on cross-sectional area alone without regard to velocity. In the United States, organized measurement of precipitation started under the Surgeon General of the Army in 1819, was transferred to the Signal Corps in 1870, and finally, in 1891, to a newly organized U. S. Weather Bureau, renamed the National Weather Service in 1970. Scattered stream-flow measurements were made on the Mississippi River as early as 1848, but a systematic program was not started until 1888, when the U. S. Geological Survey undertook this work. It is not surprising, therefore, that little quantitative work in hydrology was done before the early years of the twentieth century, when men such as Horton, Mead, and Sherman began to explore the field. The great expansion of activity in flood control, irrigation, soil conservation, and related fields which began about 1930 gave the first real impetus to organized research in hydrology, as need for more precise design data became evident. Most of today's concepts of hydrology date from 1930.

## 2  Hydrology in Engineering

Hydrology is used in engineering mainly in connection with the design and operation of hydraulic structures. What flood flows can be expected at a spillway or highway culvert or in a city drainage system? What reservoir capacity is required to assure adequate water for irrigation or municipal water supply during droughts? What effects will reservoirs, levees, and other control works exert on flood flows in a stream? These are typical of questions the hydrologist is expected to answer.

Large organizations such as federal and state water agencies can maintain staffs of hydrologic specialists to analyze their problems, but smaller offices often have insufficient hydrologic work for full-time specialists. Hence, many civil engineers are called upon for occasional hydrologic studies. It is probable that these civil engineers deal with a larger number of projects and greater annual dollar volume than the specialists do. In any event, it seems that knowledge of the fundamentals of hydrology is an essential part of the civil engineer's training.

## 3 Subject Matter of Hydrology

Hydrology deals with many topics. The subject matter as presented in this book can be broadly classified into two phases: data collection and methods of analysis. Chapters 2 to 6 deal with the basic data of hydrology. Adequate basic data are essential to any science, and hydrology is no exception. In fact, the complex features of the natural processes involved in hydrologic phenomena make it difficult to treat many hydrologic processes by rigorous deductive reasoning. One can not always start with a basic physical law and from this determine the hydrologic result to be expected. Rather, it is necessary to start with a mass of observed facts, analyze these facts, and from this analysis to establish the systematic pattern that governs these events. Thus, without adequate historical data for the particular problem area, the hydrologist is in a difficult position. Most countries have one or more government agencies with responsibility for data collection. It is important that the student learn how these data are collected and published, the limitations on their accuracy, and the proper methods of interpretation and adjustment.

Typical hydrologic problems involve estimates of extremes not observed in a small date sample, hydrologic characteristics at locations where no data have been collected (such locations are much more numerous than sites with data), or estimates of the effects of man's actions on the hydrologic characteristics of an area. Generally, each hydrologic problem is unique in that it deals with a distinct set of physical conditions within a specific river basin. Hence, quantitative conclusions of one analysis are often not directly transferable to another problem. However, the general solution for most problems can be developed from application of a few relatively basic concepts.

## New Words and Expressions

1.hydraulic *a.* 水力（学）的，水工的，水利的

2.mist *n.* 烟云，（烟，油，轻）雾

3.prehistory *n.* （历史记载以前的）史前史

4.prehistoric *a.* 历史以前的，史前的

5.pile *n.* 堆，桩

6.drought *n.* 干旱季节，旱灾

7.water *vt.* 浇水，灌溉

8.writing *n.* 书写，著作，文学作品

9.desalt *vt.* 除去…盐分

10.springs *n.*（*pl.*）根源，源泉，原动力

11.role *n.* 作用

12.astronomer *n.* 天文学家

13.pan *n.* 盘子
14.commissioner *n.* 专员
15.surgeon *n.* 外科医生，军医
16.corps *n.*（*pl.*）军团
17.scatter *vt.* 使分散
18.systematic *a.* 有系统的
19.geological *a.* 地质(学)的
20.undertake *vt.* 着手做，从事，承担，接受
21.impetus *n.* 动力，原动力
22.federal *a.* 联邦的，联合的
23.agency *n.* 机构，力量
24.specialist *n.* 专家
25.hence *ad.* 从此以后，今后，因此
26.occasional *a.* 偶然的，非经常的
27.present *vt.* 介绍；提出
28.exception *n.* 除外，例外
29.deductive *a.* 推论的
30.reasoning *n.* 推论，推理，论证
31.establish *vt.* 建立，设立，制定，规定
32.responsibility *n.* 责任
33.interpretation *n.*（资料的）整理分析，解释，阐释，阐明
34.sample *n.* 样品，试样，样本
35.unique *a.* 惟一的，独一无二的
36.distinct *a.* 与其他不同的，独特的
37.quantitative *a.* 量的，数量的，定量的
38.conclusion *n.* 结论，推论
39.transferable *a.* 可转移的，可传的
40.be lost in the mists of prehistory 史前渐被遗忘
41.ca.=circa [拉]大约
42.A. D. [拉] Anno Domini 公元
43.annual dollar volume 年费用
44.Greek 希腊人，希腊的
45.Roman 罗马人
46.Marous Vitruvius Pollion （人名）M. V.波利欧
47.Leonardo da Vinci （人名）L.达芬奇
48.Pierre Perrault （人名）P.贝罗特
49.Seine River 塞纳河
50.Halley （人名）哈雷（英国天文学家，哈雷彗星发现者）
51.Frontinus （人名）福朗堤努斯
52.U. S. Weather Bureau 美国气象局
53.the National Weather Service （美国）国家气象局
54.Mississippi River 密西西比河
55.the U. S. Geological Survey 美国地质调查局
56.Horton （人名）霍顿
57.Mead （人名）米德
58.Sherman （人名）谢尔曼
59.Surgeon General （英国陆军中的）军医；（美国）军医局局长，军医总监
60.the Signal Corps 通信兵团

## Reading Material The Runoff Cycle

The runoff cycle is the descriptive term applied to that portion of the hydrologic cycle between incident precipitation over land areas and subsequent discharge of this water through stream channels or evapotranspiration. Hoyt has presented a comprehensive description of the hydrologic phenomena occurring at selected times during the runoff cycle by considering an idealized cross section of a basin.

Fig. 3.1 shows schematically the time variations of the hydrologic factors during an extensive storm on a relatively dry basin. The dotted area of the figure represents the portion of total precipitation which eventually becomes stream-flow measured at the basin outlet. Channel precipitation is the only increment of stream-flow during the initial period of rainfall. As streams rise, surface area and consequently the volume rate of channel precipitation increase.

The rate of interception is high at the beginning of rain, especially during summer and with dense vegetal cover. However, the available storage capacity is depleted rather quickly, so that the interception rate decreases to that required to replace water evaporated from the vegetation.

The rate at which depression storage is filled also decreases rapidly from a high initial value as the smaller depressions become filled and approaches zero at a relatively high value of total-storm rainfall. Depression storage is water retained in depressions until returned to the atmosphere through evaporation.

Except in very intense storms, the greater portion of the soil moisture deficiency is satisfied before appreciable surface runoff takes place. However, some of the rain occurring late in the storm undoubtedly becomes soil moisture, since the downward movement of this water is relatively slow.

Water infiltrating the soil surface and not retained as soil moisture either moves to the stream as inter-flow or penetrates to the water table and eventually reaches the stream as groundwater. The rate of surface runoff starts at zero, increases slowly at first and then more rapidly, eventually approaching a relatively constant percentage of the rainfall rate. Both the percentage and the rate of runoff depend upon rainfall intensity.

Fig. 3.1 illustrates only one of an infinite number of possible cases. A change in rainfall intensity would change the relative magnitude of all the factors. Further complications are introduced by varying rainfall intensity during the storm or by occurrence of snow or frozen ground. To appreciate further the complexity of the process in a natural basin, remember that all the factors of Fig. 3.1 vary from point to point within the basin during a storm. Nevertheless, the foregoing description should aid in understanding the relative time variations of hydrologic phenomena which are important in considering the runoff relations discussed later in the chapter.

Statistical evidence, field observation, and logic suggest that runoff is rarely generated uniformly over a watershed. Variations in rainfall amount and intensity, soil characteristics, vegetal cover, antecedent moisture, and topography all act to create a complex pattern of behavior in which runoff from most storms derives from a relatively small portion of the watershed closest to the stream channels. Exact definition of these source areas would require vastly more detail in the data employed for rainfall-runoff estimation than is ever likely to be reasonably available, thus frustrating a wholly theoretical approach to the problem. Relative statistical constancy of the source areas appears to exist, and thus the reliability of empirically derived rainfall-runoff relations is far better than might be expected from a physical analysis of the problem.

It will be evident from Fig. 3.1 that the rainfall-runoff process is relatively complex. Despite this, the practice of estimating runoff as a fixed percentage of rainfall is the most commonly used method in design of urban storm drainage facilities, highway culverts, and many small water-control structures. The method can be correct only when dealing with a surface which is completely impervious so that the applicable runoff coefficient is 1.00.

Computer simulation techniques offer the most reliable method of computing runoff from rainfall because they permit a relatively detailed analysis using short time intervals. The type of computation used in computer simulation would be virtually impossible to carry through by hand because of the detailed computations required. The constraints of hand calculation led to methods using longer time intervals and a correspondingly less rigorous model.

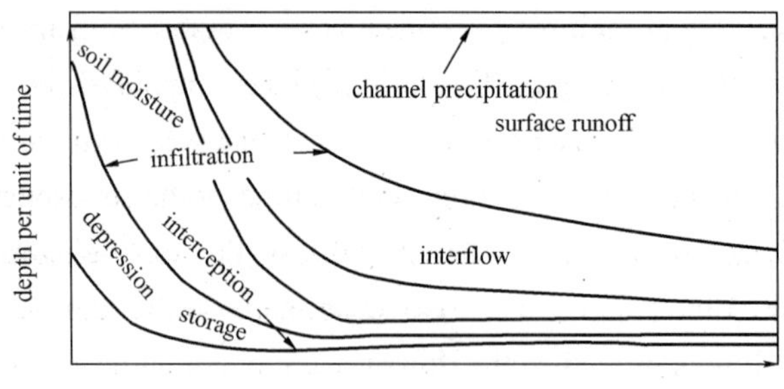


Fig.3.1 Schematic diagram of the disposition of storm rainfall

# New Words and Expressions

1.discriptive *a.* 描写的，叙述性的

2.subsequent *a.* 后成的，后起的

3.evapotranspiration *n.* 蒸散（的），总蒸发

4.idealize *vt.* （使）理想化，形成理想

5.schematically *ad.* 用示意图

6.dotted *a.* 有点的

7.streamflow *n.* 河道水流，河道流量

8.increment *n.* 增量

9.interception *n.* 截留

10.vegetal *a.* 植物的

11.deplete *vt.* 使枯竭，放空；减少，用尽，消耗

12.vegetation *n.* (植物)生长，植被，植物

13.depression *n.* 洼地；降低，减少

14.deficiency *n.* 缺乏，不足

15.appreciable *a.* 明显的，相当大的

16.infiltrate *vt.* 渗入，渗透

17.penetrate *vi.* 渗入，渗透

18.interflow *n.* 壤中流，土内水流

19.infinite *a.* 无数的，不定的

20.foregoing *a.* 前面的，上述的

21.statistical *a.* 统计的

22.watershed *n.* 集水区，流域

23.antecedent *a.* 前期的，起初的

24.frustrat *vt.* 破坏，阻止；使落空

25.constancy *n.* 不变，恒定性，稳定性

26.empirically *ad.* 经验上地，以实验为根据地

27.culvert *n.* 涵洞，排水渠，下水道

28.coefficient *n.* 系数

29.simulation *n.* 模拟

30.virtually *ad.* 实际上

31.constraint *n.* 约束

32.correspondingly *ad.* 相当地，合适地

33.rigorous *a.* 严格的，精密的

34.precipitation 降雨量

35.subsequent discharge 后成排泄量

36.channel precipitation 河面降水，河面降雨量

37.depression storage 洼地蓄水量，洼地容积，填洼量

38.water table 地下水面；地下水位

39.rainfall intensity 降雨强度，雨强

40.antecedent moisture 前期含水量

# Lesson 4 Underground Water

Of all the earth's water, 97% is found in the oceans, 2% in glaciers and only 1% on land. Of this 1% almost all (97%) is found beneath the surface and called sub-surface or underground water. Most of this water eventually finds its way back to the sea either by underground movement or by rising into surface streams and lakes.

These vast underground water deposits provide much needed moisture for dry areas and irrigated districts. Underground water acts in similar ways to surface water, also performing geomorphic work as an agent of gradation.

Even though man has been aware of sub-surface water since earliest times, its nature, occurrence, movement and geomorphic significance have remained obscure. Recently, however, some answers have been found to the perplexing questions about underground water's relationship to the hydrological cycle.

## 1 Source of Underground Water

Since the days of Vitruvius at the time of Christ, many theories have been presented to explain the large volume of water underneath the earth's surface. One theory was that only the sea could provide such large quantities, the water moving underground from coastal areas. Vitruvius was the first to recognize that precipitation provided the main source of sub-surface water, although his explanations of the mechanics involved were not very scientific.

His theory, now firmly established, is termed the infiltration theory, and states that underground water is the result of water seeping downwards from the surface, either directly from precipitation or indirectly from streams and lakes. This form of water is termed meteoric. A very small proportion of the total volume of sub-surface water is derived from other sources. Connate water is that which is trapped in sedimentary beds during their time of formation. Juvenile water is water added to the crust by diastrophic causes at a considerable depth, an example being volcanic water.

## 2 Distribution of Sub-Surface Water

During precipitation water infiltrates into the ground, under the influence of gravity, this water travels downwards through the minute pore spaces between the soil particles until it reaches a layer of impervious bedrock, through which it cannot penetrate. The excess moisture draining downwards then fills up all the pore spaces between the soil particles, displacing the soil air. During times of excessive rainfall such saturated soil may be found throughout the soil profile, while during periods of drought it may be non-existent. Normally the upper limit of saturated soil, termed the water table, is a meter or so below the surface, the height depending on soil characteristics and rainfall supply.

According to the degree of water-occupied pore space, sub-surface moisture is divided into two zones: the zone of aeration and the zone of saturation, as illustrated in Fig. 4.1.

### 2.1 Zone of Aeration

This area extends from the surface down to the upper level of saturation—the water table. With respect to the occurrence and circulation of the water contained in it, this zone can be further divided into three belts: the soil water belt, the intermediate belt and the capillary fringe (Fig.4.1).

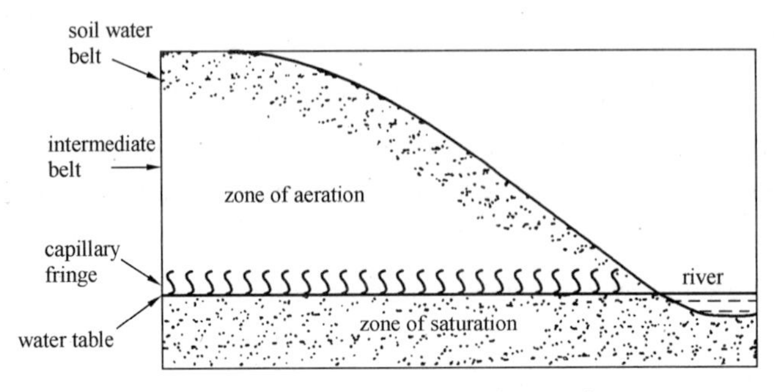


Fig. 4.1 Zone of aeration and saturation

**(1) Soil Water Belt.** Assuming that the soil is dry, initial rainfall allows water to infiltrate, the amount of infiltration depending on the soil structure. Soils composed mainly of large particles, with large pore spaces between each particle, normally experience a more rapid rate of infiltration than do soils composed of minute particles. No matter what the soil is composed of some water is held on the soil particles as a surface film by molecular attraction, resisting gravitational movement downwards. The water held in this manner is referred to as hygroscopic water. Even though it is not affected by gravity, it can be evaporated, though not normally taken up by plants.

**(2) Intermediate Belt.** This belt occurs during dry periods when the water table is at considerable depth below the surface. It is similar to the soil water belt in that the water is held on the soil particles by molecular attraction, but differs in that the films of moisture are not available for transpiration or for evaporation back to the atmosphere. In humid areas, with a fairly reliable rainfall, this belt may be non-existent or very shallow. Through it, gravitational or vadose water drips downwards to the zone of saturation.

**(3) Capillary Fringe.** Immediately above the water table is a very shallow zone of water which has been drawn upwards from the ground-water reservoir below by capillary force. The depth of this zone depends entirely on soil texture, soils with minute pore spaces being able to attract more water from below than soils with large pore spaces. In the latter types of soil the molecular forces are not able to span the gaps between soil particles. Thus, sandy soils seldom exhibit an extensive capillary fringe, merging from soil water through to the zone of saturation.

### 2.2 Zone of Saturation

The zone of saturation is the area of soil and rock whose pore spaces are completely filled with water, and which is entirely devoid of soil air. This zone is technically termed ground water even though the term broadly includes water in the zone of aeration. The upper limit of the zone of saturation is the water table or phreatic surface. It is difficult to know how deep the ground-water zone extends. Although most ground water is found in the upper 3 km of the crust, pore spaces capable of water retention extend to a depth of 16 km. This appears to be the upper limit of the zone of rock flowage where pressures are so great that they close any interstitial spaces.

The upper level of the saturated zone can be completely plotted by digging wells at various places. Studies suggest two quite interesting points (Fig.4.2).

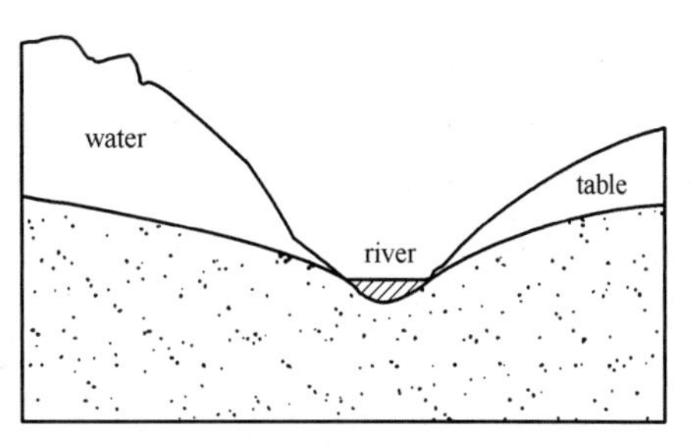

Fig. 4.2 Situation of the water table

1) The water table level is highest under the highest parts of the surface, and lowest under the lowest parts of the surface. Hills and mountains have a higher level phreatic surface than valleys and lakes. The reason for this is that water continually percolating through the zone of aeration lifts the water table, while seepage from the ground-water zone into creeks and lakes lowers the level.

2) The depth of the water table below the land surface is greatest in upland areas where the water moves quite freely downhill under gravity. Close to streams, lakes and swamps the water table is close to, if not at the surface, as water from the higher areas builds it up.

**(to be continued)**

## New Words and Expressions

1.geomorphic *a.* 地貌的
2.agent *n.* （发生作用或影响的）动因，力量
3.gradation *n.* 均夷作用；分等，分级；粒级（分粒）作用
4.obscure *a.* 不清楚的，不引人注目的
5.perplexing *a.* 使人困惑的，错综复杂的
6.meteoric *a.* 大气的，气象（学）的
7.connate *a.* 同源的，同生的
8.juvenile *a.* 岩浆的
9.diastrophic *a.* 地壳运动的
10.volcanic *a.* 火山（性）的；猛烈的
11.minute *a.* 细小的，微小的
12.bedrock *n.* 基岩
13.aeration *n.* 通风，通气
14.capillary *n.* 毛细作用的，毛细现象的
15.fringe *n.* 边缘，端
16.hygroscopic *a.* 吸湿的，收湿的
17.film *n.* 薄层；膜，薄膜
18.vadose *a.* 渗流
19.phreatic *a.* 井的
20.flowage *n.* 泛滥，积水
21.interstitial *a.* 空隙的
22.upland *a.* 高地的，山地的
23.downhill *n.* 下坡
24.sub-surface water 地下水
25.Vitruvius 维特鲁维亚（罗马建筑师）

## Reading Material Underground water

**(continue)**

### 3 Quantity and Holding Capacity

The quantity of water capable of being held in the ground can be determined by knowledge of several factors.

### 3.1 Porosity

The porosity of a particular rock means the amount of pore space or interstitial space in a given volume. In saturated bedrock, the porosity is expressed by the percentage of the total volume of rock that is occupied by water. With regard to the different degrees of porosity of sedimentary rocks, it is interesting to note that both sandstone and mudstone (silt and clay) have high porosities, the former having fewer, but much bigger, pore spaces than the latter. In igneous or primary rocks porosity is particularly low, there being few pore spaces between each mineral grain of granite or other massive rocks. Basalt, particularly scoriaceous basalt, may be quite porous as a result of heavy jointing or of gas bubbles which formed as the lava cooled.

### 3.2 Permeability

If the pore spaces between individual particles are connected, the water is capable of movement, both laterally and under gravity. Rocks permitting this movement are termed permeable. Rocks which preclude movement of water through them, either because of blocked interstitial spaces or spaces so minute that water cannot travel through them, are termed impermeable.

Rocks comprising coarse sands and gravels allow quite rapid water movement, whereas finer clay and silt rocks prevent movement. In the latter cases pore spaces are so small that the water held on each particle by molecular attraction effectively takes up all the pore space. There is a far greater expanse of surface area when then the particles are minute, so that more moisture can adhere by molecular attraction. Most of the primary igneous rocks are impermeable unless heavily jointed.

Thus, although sandstone and mudstone are both quite porous, sandstone is permeable while mudstone is impermeable.

### 3.3 Specific Yield and Retention

Although somewhat similar in concept to both porosity and permeability, the specific yield of a rock means the amount of free water capable of movement and the specific retention means the amount of dead water, or water trapped in the rock mass. Both of these are stated as percentages of the total volume of rock.

In coarse sand the specific retention is low and the specific yield high. Fine clay has a high retention rate and a low specific yield, even though its porosity is quite high.

### 3.4 Rock Strata

As mentioned in previous paragraphs, rock types differ widely in their degree of porosity and permeability. Rocks which are both porous and permeable allow vast quantities of water to move quite freely. These are known as aquifers, common examples being sandstone and grit. Rocks which prohibit movement, even though they may be highly porous, are termed aquicludes, the best examples being mudstone and shale.

In the case of Fig. 4.3(a) the horizontal arrangement of sandstone and shale restricts the amount of water that can be absorbed below the surface. If, as in Fig. 4.3(b), these beds were inclined to the surface, a greater extent of permeable sandstone would allow an increased amount of water to infiltrate, and the depth to which this water could penetrate would be increased.

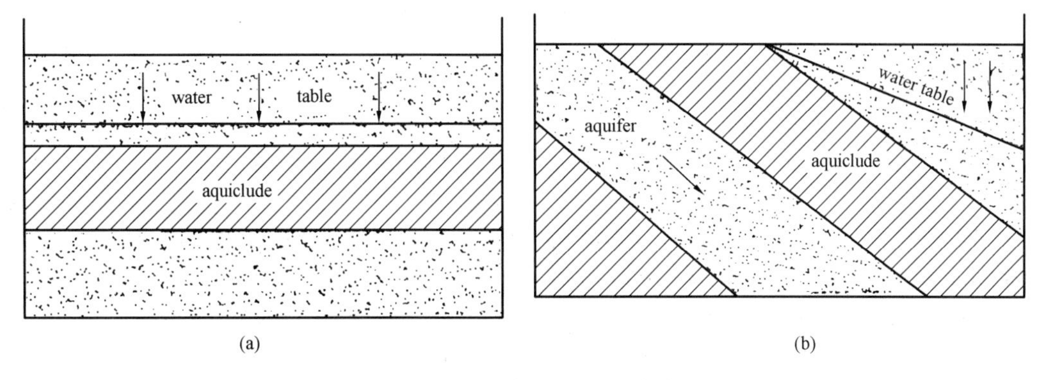


Fig. 4.3 Aquifers and aquicludes

(a) precipitation; (b) precipitation

Any aquiclude prohibiting the downward movement of surface moisture may cause a superficial water table, one that is considerably higher than the main water table below. This formation is referred to as a perched water table (Fig. 4.4).

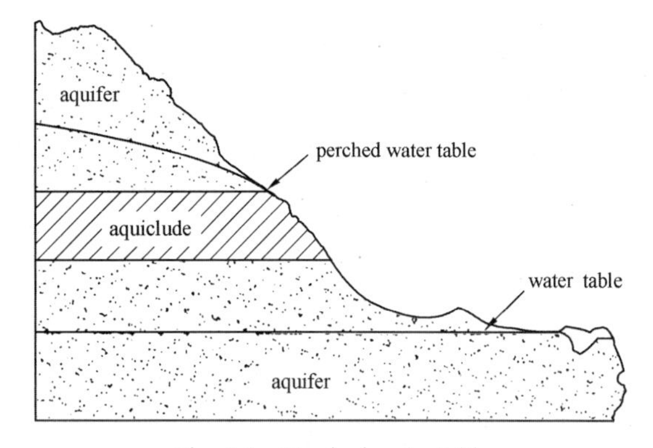


Fig. 4.4 Perched water table

### 3.5 Precipitation

As precipitation is the primary source of all ground water, it is logical to expect that variations in quantity or distribution will cause fluctuations in the ground-water level.

The upper level of the ground-water zone is generally closer to the surface in high rainfall regions than in arid areas, since there is more water available for infiltration.

Uniform rain maintains the hygroscopic water and allows all excess to drain downwards to the zone of saturation. In drier areas where rainfall is particularly variable, the hygroscopic water is rapidly evaporated, so that any rainfall must first of all satisfy the demands of individual soil particles in the zone of aeration before infiltrating farther. In humid areas experiencing a low rainfall intensity (less rain per wet day) the amount of water infiltrating into the soil and rock is larger than in areas with a high intensity. The movement of the water table according to both seasonal and periodical fluctuations of rainfall can be seen in Fig. 4.5.

### 3.6 Evaporation

As we have seen in earlier discussions, Areas experiencing a high evaporation rate are less likely to have large quantities of soil water, as much of the infiltrating moisture will be returned immediately to the atmosphere. Areas with a moderate amount and intensity of rainfall are more likely to have greater quantities of water infiltrated.

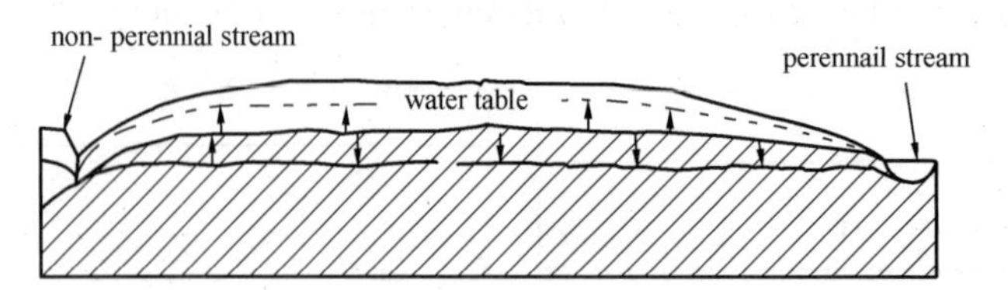


Fig. 4.5 Seasonal fluctuation in the height of water table

If the water table is close to the surface and the soil above it close to maximum capacity, soil evaporation may occur. The rate of such evaporation decreases very quickly with increasing distance of the water table from the surface and drops to zero below the upper limit of capillary rise. In sand the capillary rise is about 0.6 meter, while in clay soils it may be as high as 3 meters.

### 3.7 Slope

As we have seen in earlier discussions of runoff, runoff increases with the angle of slope. Infiltration to the zone of saturation is negligible on steep slopes compared with low-angle slopes where the runoff is not as rapid, and there is more time for the same quantity of moisture to infiltrate.

### 3.8 Vegetation

Vegetational effects on the amount of water held under the surface are rather indirect. The root system of plants and their surface extent retard runoff, which increases infiltration. However, if the rainfall total is mediocre and the land surface is flat, vegetation may utilize much of the infiltrating water for its growth, returning it to the atmosphere by transpiration.

## 4 Movement of Ground Water

Unlike the movement of water in surface channels, water passing through the crust moves differently and very slowly. The flow in rivers and creeks, because of their relatively high velocities, is generally turbulent. However, the water passing through interstitial spaces in saturated material is characteristically laminar or streamlined, with individual particles of water following roughly parallel paths. This form of movement is termed percolation.

The rate of movement of each thread of water varies considerably. Those nearest the walls of the interstitial openings are slowest, and those in the middle are the most rapid. The rate of movement of any individual particle also differs: when approaching a narrow neck in the pore space, the particle speeds up and coalesces with other threads, and then separates and slows down in the larger opening.

From the time water infiltrates into the ground at any point in the intake area, its mode of flow is governed by some point where water is being discharged. Normally, this would be a surface river in its valley. Contrary to what might be expected, the path water takes is never a straight line from intake to discharge area. Rather, as in Fig.4.6, the particles travel elliptical paths, varying both in distance traveled and in depth below the water table. The mechanics of this movement are simple. The long curved paths are due to the tendency of water to move downwards under gravity and laterally according to the slope of the water table. The difference in the level of the water table, called the hydraulic head, is responsible for a pressure build up which dictates the paths of flow. Where there is unequal pressure, water will flow from an area of high pressure to an area of low pressure.

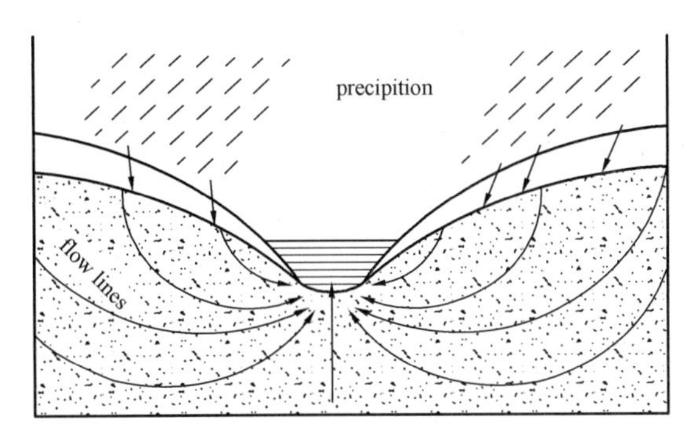


Fig.4.6 Movement of underground water

## New Words and Expressions

1.porosity *n.* 孔隙度（率）多孔性，多孔结构

2.sandstone *n.* [地]沙岩

3.mudstone *n.* [地]泥岩

4.igneous *a.* 火成的

5.granite *n.* 花岗岩，花岗石

6.basalt *n.* [地]玄武岩

7.scoriaceous *a.* 火山岩矿渣的

8.porous *a.* 多孔的；有气孔的，能渗透的

9.bubble *n.* 泡，水泡，气泡

10.lava *n.* 熔岩

11.preclude *vt.* 阻碍，排除，消除

12.yield *n.* 出水率，单位出水量

13.retention *n.* 持水量，滞留量

14.grit *n.* 粗砂，砂砾

15.aquiclude *n.* 隔水层

16.shale *n.* 页岩

17.srperficial *a.* 地面的，外部的

18.perch *vt.* 位于高处

19.moderate *a.* 中等的，适度的；*vt.* 使和缓

20.thread *n.* 线，细细的一条

21.neck *n.* 狭窄地带，隘口，海峡

22.coalesce *vi.* 聚结，凝聚，结合

23.contrary *a.* 相反的，相对的

24.unequal *a.* 不相等的，不同的

25.juvenile water *n.* 岩浆（原生，初生）水

26.be devoid of 没有…的；无…的

27.igneous rock 火成岩

28.specific yield and retention 单位出水量和持水量

29.perched water 上层滞水

# Lesson 5 What Causes Flooding

What causes flooding? The basic cause is excessive runoff from catchments into river systems incapable of carrying this extra volume. Can science and technology prevent flooding or, at least, reduce its severity? Unfortunately, this is a complex problem to which as yet there is no very satisfactory solution.

Let us consider first the reduction of runoff from catchment areas. Some regions have soils which have low absorbing capacity. In a heavy rainstorm such soil is quickly saturated and all additional rainfall then runs off into the river. A seasonal variable is the moisture status of the soil at the commencement of a rainstorm. If the soil is already moist, a relatively minor storm could still cause heavy runoff because the soil is incapable of retaining additional moisture. These factors are not easily influenced by man. However, man's utilization of the catchment area can have an important influence on flooding. Large scale clearing of trees and scrub greatly reduces the capacity of the soil to retain water. It also tends to cause soil erosion which aggravates flooding by choking rivers and streams with deposited silt. Correct management of catchment areas is therefore one important approach to the problem of flood control.

A more direct approach which is used in an emergency is the construction of levees. When rising floodwaters threaten a township the citizens form work-parties to build barricades of sandbags along the river bank, hoping that these barricades will hold back the flood waters until the emergency passes. It may be wondered why levees are not usually built as permanent structures to which the town is protected at all times. The reason is that levees are an unsatisfactory solution to the problem. If a levee collapses, the floodwaters escape as a sudden deluge with increased capacity for destruction. Levees as they divert the floodwater from one area frequently create or aggravate problems in another. They can be a cause of enmity between communities for this reason.

Another approach is the construction of dams so that floodwaters can be retained in a reservoir until the crisis is over; slow release of the water during the succeeding weeks or months would then be possible. Combined purpose irrigation and flood control dams would seem to be a logical solution. Unfortunately, a reservoir which is to be used for irrigation needs to be kept nearly full in winter, while one which is to be used for flood control needs to be kept empty, so that it is available as a water store when needed. This conflict of operating requirements means that combined purpose dams are rarely feasible. Separate dams would be required for flood control and their very high cost makes this an impractical solution. The next approach to the problem is that of improving the capacity of the river to carry larger volumes of water without overflowing its banks. A number of measures are available, some simple, some complex. They all have widespread effects on the river so any of these measures should be used as part of a comprehensive plan. Work of this kind is known as "river improvement" or "river management".

One simple, but important step is to ensure that the water course of a river is kept free of obstructions. These frequently consist of dead trees which have fallen into the river, where they remain to impede the flow of water. They are called "snags" and the removal work "snagging". Many of the trees that line Australian River banks are hardwoods, which are too heavy to float so they remain where they fall. Furthermore, hardwoods are very durable; large red gum logs have been known to survive over a hundred years under water.

Another method of increasing the capacity of the river is to remove choking plant growth. Early setters introduced willow trees to many of our rivers, partly for shade, partly to recall old England and hopefully to reduce the erosion of the river banks. Unfortunately, these trees are difficult to control and willow infestation is now quite commonly a problem. Protection of the banks of a river from erosion by the stream of water is another measure. Rivers which follow a meandering, or winding course tend to erode their banks along the outer curves. This can mean a loss of valuable soil from the eroded bank area and is also a cause of local flooding. Means of protecting banks from erosion have been devised.

The simplest device used for this purpose is that of anchored or tied tree trunks along the eroded bank. The trunks protect the bank and encourage the deposition of silt on the bank so that it is gradually built up.

## New Words and Expressions

1.flooding *n.* 洪水泛滥；漫灌

2.catchment *n.* 集水，汇水；集水处；流域；集水量

3.severity *n.* 严重；危害

4.variable *a.* 易（常）的，无常的；变量

5.status *n.* 情况，状况（态）

6.commencement *n.* 开始，开端

7.utilization *n.* 利用，效用

8.clearing *n.* 砍伐；（森林中的）空旷地

9.scrub *n.* 灌木，丛林

10.erosion *n.* 侵蚀；腐蚀；侵害

11.aggravate *vt.* 使恶化，加重（剧）

12.choke *vt.* 淤塞，阻（堵）塞，塞满

13.emergency *n.* 紧急，应急

14.levee *n.* 堤；*vt.* 筑堤

15.township *n.* 镇（区），区

16.work-party *n.* 抢险队；工作组

17.barricade *n.* 路障；挡墙

18.sandbag *n.* 沙袋

19.escape *v.* 涌出，逸出

20.survive *vi.* 残存，继续存在

21.forthermore *ad.* 而且，此外

22.deluge *n.* 洪水，大水灾；大雨，暴雨

23.divert *vt.* 使转向；分水

24.inmity *n.* 不和，纠纷，敌对

25.community *n.* 居住区；地区

26.crisis （*pl.* crises） *n.* 危机，紧急关头，危险期

27.succeeding *a.* 接连的，随后的，以后的

28.feasible *a.* 可行的，有可能的

29.widespread *a.* 普遍的

30.obstuction *n.* 阻塞；障碍（物）

31.impede *vt.* 妨碍；阻止

32.snag *n.* 水中隐树，沉木；清除隐患（暗礁沉木，树根等）

33.hardwood *n.* 硬木

34.durable *a.* 耐久的，坚牢（固）的

35.gum *n.* 桉树属；橡胶

36. settler *n.* 开拓者，移居者
37.introduce *vt.* 引种
38.willow *n.* 柳树
39.recall *vt.* 收回，撤消，回顾，回忆，想起
40.infestation *n.* 蔓延；侵扰
41.meandering *a.* 蜿蜒的，曲流的
42.winding *a.* 曲折的，弯曲的，迂回的；线卷，绕组
43.erode *vt.* 侵蚀，冲刷，腐蚀
44.anchor *vt.* 锚定；固定
45.devise *vt.* 想出；设计
46.device *n.* 方法，手段
47.trunk *n.* （树干）；柱身
48.encourage *vt.* 促使
49.deposition *n.* 沉（淤）积；沉淀物
50.as yet 到目前为止仍…，现在还…
51.have an influence on… 对…有影响
52.soil erosion 土壤侵蚀
53.in an emergency 遇到紧急情况时，万一发生事故时
54.the outer curve 凹岸
55.build up 加固

# Reading Material Flood

The account of the Deluge in the Bible and similar traditional accounts are evidence that there always have been floods large enough to affect man's struggle to win a living from the earth. The early civilizations that grew along the Nile, Indus, and the Yellow Rivers are examples of the development of civilizations in an environment where a river enforced community action for flood protection. Such an environment nurtured man and his crops, but the river was also a source of disaster.

In China, the Yellow River has been the cause of repeated floods during 4,000 years of continuous settlement. The land subject to flooding has an area of about 55,000 square miles. Because of its devastating floods, the Yellow River is known as "China's Sorrow". Despite the repeated setbacks caused by the floods, however, the Chinese did not abandon their lands. This generally was true of the other great early civilizations that experienced river floods.

Most of the world's population and property are located on lands subject to the overflow of rivers or seas. For example, flood-prone lands comprise about 5% of the area of the United States, more than 10% of the Yellow River basin in China, and almost all of the Netherlands.

One major type of flood is a river flood. Many cities, including Paris, Rome, and Washington occupy land subject to river floods.

The flow of a river usually is confined to a well-defined channel that meanders in the course of time from one side of its valley to the other. However, when there is a heavy rainfall or rapid snowmelt, the river overflows into a wide flat area adjacent to the channel. This area, the natural floodway of the river, is called a floodplain, which is composed of sediments deposited by the river. Typically, a river uses some portion of its floodplain about once in two or three years. Once in a century a river may inundate its entire floodplain to a considerable depth.

A second major type of flood is a coastal flood. Coastal lands, such as offshore bars formed by sediments carried by coastal currents occupy a position relative to the sea that floodplains do to rivers. Like floodplains, offshore bars and barrier beaches are the sites of many cities—for example, Atlantic City, Lagos and Nigeria. All are subject to coastal floods.

The floodplains, the coastal bars and the ocean strands invite man's occupancy. Floodplain soils normally are more fertile and easier to till than uplands, and the flatlands are less costly to build on. Because rivers and oceans have been the main avenues of commerce cities naturally were built on the contiguous lands. For these reasons, overflow of rivers or the sea destroys crops, causes extensive damage to property, cuts transportation lines, disrupts the life of cities, and takes a toll in lives. Loss of life as a consequence of floods is decreasing in industrial countries that use protective and flood-warning measures, but its toll continues unabated in less developed countries.

In the United States, the floodplains of the Ohio River and its major tributaries have been the scene of the greatest flood damages. The industrial Northeast and Middle Atlantic regions come a close second. No region is spared from flood damages, not even the deserts, where sporadic floods form fearsome walls of water that rush down normally dry channels and wash out roads and bridges.

## New Words and Expressions

1.deluge　*n.* 洪水，大雨
2.Bible　*n.* （基督教的）《圣经》
3.enforce　*v.* 实施
4.nurture　*vt.* 养育，培育
5.devastate　*vt.* 破坏，毁坏
6.setback　*n.* 挫折
7.occurrence　*n.* 发生[现]，出现（率）
8.overflow　*vt.* 使溢出，使泛滥
　overflow　*v., n.* 溢流，泄出，泛滥
9.comprise　*vt.* 构成，包括
10.flood-prone　*n.* 洪水侵袭
11.well-defined　*a.* 固定的，清楚的
12.meander　*vi.* 蜿蜒，曲折地流
13.floodway　*n.* 泄洪道，分洪道
14.inundate　*vt.* 淹没，泛滥
15.offshore　*a.* 近海（的），向海（的）
　offshore beach　近海（沙）洲
16.barrier　*n.* 堤，障碍物
　barrier beach　滨外滩，滨外沙埂
17.strand　*n.* 沙岸，（海、湖）滨
18.occupancy　*n.* 占用
19.till　*v.* 耕作
20.upland　*n.* 高地
21.contiguous　*a.* 邻接的，连接的
22.disrupt　*vt.*毁坏，中断
23.toll　*n.* 损失，牺牲
24.consequence　*n.* 结果，后果
25.unabated　*a.* 未降低的，未减轻的
26.tributary　*n.* 支流
27.scene　*n.* 现场，出事
28.spare　*v.* 不损害
29.sporadic　*a.* 偶尔发生的，分散的，零星的
30.fearsome　*a.* 可怕的
31.Indus　*n.* 印度河
32.Ntherlands　*n.* 荷兰
33.Atlantic City　大西洋城（美国新泽西州东南部城市）
34.Lagos　*n.* 拉各斯（尼日利亚首都）
35.Nigeria　*n.* 尼日利亚
36.the Ohio River　俄亥俄河
37.be located on　位于…上
38.be confined to　（被）限制在，（被）封闭在
39.in the course time　终于，最后，经过一定时间
40.(be) relative to　关于，和…有关；相对于
41.take (a) toll of　使遭受损失（伤亡）
42.as a consequence of　由于…的结果，因为…（的缘故）
43.wash out　冲掉，冲坏

# Lesson 6 Nature of Water Pollution

Water, one of man's most precious resources, is generally taken for granted until its use is threatened by reduced availability or quality. Water pollution is produced primarily by the activities of man, specifically his mismanagement of water resources. The pollutants are any chemical, physical, or biological substances that affect the natural condition of water or its intended use. Because water pollution threatens the availability, quality and usefulness of water, it is of worldwide critical concern.

The increase in the number and variety of uses for water throughout the world has produced a wide range of standards of water quality that must be satisfied. These demands include: ① preservation of rivers in their natural state; ② potability of the water supply; ③ preservation and enhancement of fish and wildlife; ④ safety for agricultural use; ⑤ safety for recreational use, including swimming; ⑥ accommodation to a great variety of industrial purpose; ⑦ freedom from nuisance; ⑧ generation of power for public utilities; ⑨ dilution and transport of wastes. Besides the specific chemical, biological, and physical requirements for the multitude of uses noted above, there are constraints reflecting public health requirements, aesthetics, economics, and short and long-term ecological impacts. Consequently, there is no rigid or specific definition of water pollution, since the intended use or uses of the water must be taken into consideration in any definition of what constitutes polluted water.

One method of classifying the gaseous, liquid and solid constituents of water that constitute pollution depends on the intended use of the water. The pollutants are then grouped as not permissible, as undesirable and objectionable, as permissible but not necessarily desirable, or as desirable. For example, if water is to be used immediately for animal consumption, toxic compounds are not desirable, whereas a certain amount of oxygen is not objectionable. On the other hand, if the water is to be used in a power plant for steam generation, toxic materials might be allowable or even perhaps desirable, whereas oxygen that could possibly corrode equipment would be objectionable.

Another method of classifying pollutants that enter water as a result of man's domestic, industrial or other activities is to distinguish between conservative and non-conservative pollutants. Conservative pollutants are those that are not altered by the biological processes occurring in natural waters. These pollutants are for the most part inorganic chemicals, which are diluted in receiving water but are not appreciably changed in total quantity. Industrial wastes contain numerous such pollutants, including metallic salts and other toxic, corrosive, colored, and taste-producing materials. Domestic pollution and return flow from irrigation may contain numerous such pollutants, including chlorides and nitrates.

Non-conservative pollutants, on the other hand, are changed in form of reduced in quantity by chemical and physical processes involved in biological phenomena occurring in water. The most common source of non-conservative pollutants is domestic sewage—highly putrescible organic waste that can be converted into inorganic materials such as bicarbonates, sulfates, and phosphates by the bacteria and other microorganism in the water.

If the water is not too heavily laden with wastes, it will undergo "self-purification". This process involves the action of aerobic bacteria, that is, bacteria that require free oxygen to break down wastes, and it produces no offensive odors.

If, however, the water is laden with wastes beyond a certain amount, the process of biological degradation becomes anaerobic. That is, it proceeds by the action of bacteria that do not require free oxygen. In the process, noxious hydrogen sulphide gas, methane, and other gases are produced. The aerobic and anaerobic processes that occur naturally in streams are used in sewage treatment plants and are, in fact, major elements in sewage treatment.

The problem of water pollution has been and is almost worldwide.

## New Words and Expressions

1.availability *n.* 可得性，可用性
2.potability *n.* 饮用
3.enhancement *n.* 增加，提高
4.accommodation *n.* 适应；供应
5.nuisance *n.* 有害的东西
6.utility *n.* 公用事业
7.multitude *n.* 多，许多
8.aesthetics *n.* 美学
9.ecological *a.* 生态学的
10.rigid *a.* 严格的
11.constitute *vt.* 构成，形成
12.group *vt.* 分组，分类
13.objectionable *a.* 不适合的
14.conservative *a.* 保守的，不变的
15.alter *v.* 改变，变化
16.dilute *v.* 稀释，冲淡
17.appreciably *ad.* 明显地，可观地
18.nitrate *n.* 硝酸盐
19.putrescible *a.* 会腐败的，容易腐烂的
20.bicarbonate *n.* 重碳酸盐
21.sulfate *n.* 硫酸盐
22.phosphate *n.* 磷酸盐
23.aerobic *a.* 需氧的，有氧的
24.offensive *a.* 讨厌的，令人不愉快的
25.degradation *n.* 降解，减低
26.noxious *a.* 有害的，有毒的
27.take … for granted 认为…是理所当然的；对…不当一回事
28.freedom from 免除
29.a multitude of 许多的，众多的
30.take … into consideration 考虑到，注意到
31.distinguish between 区别…和…
32.for the most part 大概，多半
33.be about to do sth. 即将（做），（不久）就要（做）
34.be laden with 装满

## Reading Material The History of Water Pollution

By the 20th century, people had become concerned about water pollution, largely because of the prevalence of waterborne diseases such as typhoid fever. The coliform test, still the major indicator of the biological safety of water, was developed to detect the presence or absence in water of bacterial organisms from the human intestinal tract. Measures were aimed also at controlling water temperature. Other measures were developed to attack organisms, to make use of the natural assimilation capacity of flowing streams in purifying wastes, and to pass water through treatment plants to provide safe drinking water. Later, pollution control was directed toward problems caused by processing wastes from industry. Control was aimed primarily at protecting downstream public water supplies or at stopping or preventing public nuisances.

## 1 Factors Affecting Water Pollution in the Mid-20th Century

The growth and changes in many parts of the world, particularly in the more developed nations, in the mid-20th century have contributed substantially to water pollution. Population growth, the rate of industrial progress and technological developments, changing land practices, and other factors have accelerated water pollution—largely through the need for waste disposal.

Rapidly growing urban areas have created demand for disposal of domestic wastes through already overburdened waterborne collection and disposal systems. The increased production of goods has greatly increased the amount of common industrial wastes. New processes in manufacturing produce new and complex wastes that are not removed by existing treatment and control technology. Increased use of commercial fertilizers and the development of new pesticides applied to agricultural and urban areas have created a host of new pollution problems. Expansion in the nuclear energy field and the use of radioactive materials has presented new and complex water pollution problems. Leapfrogging urban developments have made it difficult to construct and extend well-designed sewer systems and to develop treatment plants of adequate size. The intrusion of seawater in response to excessive pumping of groundwater supplies has created new problems of groundwater degradation.

## 2 Present Conditions

Pollution is a problem in almost all waterways—rivers, lakes, estuaries, and even oceans, once erroneously thought to be able to assimilate almost everything.

The seriousness of the water pollution problem can be illustrated by a few examples in the United States. In the mid of 1960's the Arkansas River was found routinely to have large amounts of sodium in the water, making it unusable for drinking. A 1964 outbreak of encephalitis in New Jersey was attributed in part to water pollution since polluted streams provided extensive breeding grounds for the mosquito carriers of the disease. As late as 1965 a Midwestern city was dumping 300,000 pounds of paunch manure (partly digested material in the stomachs of slaughtered cattle) into the Missouri River each day. This material contributed to the unsightliness of the river, led to its condemnation for recreation and sports, and of course, presented public health hazards.

# New Words and Expressions

1.coliform *n.* 大肠杆菌

2.estuary *n.* 河口，江（海），港（湾）

3.erroneous *a.* 错误的，不正确的

4.assimilate *n.* （使）同化，吸收，消化

5.encephalitis *n.*（*pl.* encephalitides）脑炎，大脑炎

6.outbreak *n.*，*vt.* 爆发，反抗，冲破

7.dump *v.* 排泄，倾倒，抛弃

8.slaughter *n.*，*vt.* 屠杀，屠宰

# Lesson 7 Planning for Water Resources Development

Planning can be defined as the orderly consideration of a project from the original statement of purpose through the evaluation of alternatives to the final decision on a course of action. It includes all the work associated with the design of a project except the detailed engineering of the structures. It is the basis for the decision to proceed with (or to abandon) a proposed project and is the most important aspect of the engineering for the project. Because each water-development project is unique in its physical and economic setting, it is impossible to describe a simple process that will inevitably lead to the best decision. There is no substitute for "engineering judgment" in the selection of the method of approach to project planning, but each individual step toward the final decision should be supported by quantitative analysis rather than estimates or judgment whenever possible. One often hears the phrase "river-basin planning", but the planning phase is no less important in the case of the smallest project. The planning for an entire river basin involves a much more complex planning effort than the single project, but the difficulties in arriving at the correct decision may be just as great for the individual project.

The term "planning" carries another connotation which is different from the meaning described above. This is the concept of the regional master plan which attempts to define the most desirable future growth pattern for an area. If the master plan is in reality the most desirable pattern of development, then future growth should be guided toward this pattern. Unfortunately, the concept of "most desirable" is subjective, and it is difficult to assure that any master plan meets this high standard when first developed. Subsequent changes in technology, economic development, and public attitude often make a master plan obsolete in a relatively short time. Any plan is based on assumptions regarding the future, and if these assumptions are not realized the plan must be revised. Plans generally must be revised periodically.

An overall regional water-management plan, developed with care and closely coordinated with other regional plans, may be a useful tool in determining which of many possible actions should be taken. But it must always be considered subject to modification as the technological, economic and social environment change or as new factual data that are developed. A master plan that is no more than a catalog of all physically feasible actions is likely to prove of little value.

Planning occurs at many levels within each country with the purpose and nature of the planning effort differing at each level. Many countries have a national planning organization with the goal of enhancing the economic growth and social conditions within the country. Even if no such organization exists, the national goals remain, and some form of national planning occurs in the legislative or executive branches of government. The national planning organization will rarely deal with water problems directly, but in setting goals for production of food, energy, industrial goods, housing, etc, it may effectively specify targets for water management.

To allow for differences between the various regions of the country, regional planning groups may exist and equivalent regional water planning may occur. However, a natural "region" for water planning is the river basin and the tendency is to create river basin commissions, or river basin management units (authorities). These groups must assure coordination between the various activities within the river basin. Each specific action in water management is likely to have consequences downstream (and sometimes upstream), and thus these specific actions should not be planned in isolation but must be coordinated.

Planning of specific actions is the lowest level of planning, but it is at this level that important decisions which determine the effectiveness of water management are made. This level is often called project planning although a physical project may not necessarily result. For example, a study leading to a plan for flood-plain management is a legitimate project plan.

Project planning usually passes through several phases before the final plan emerges. In each country there is a specified sequence with specific names. The first phase or reconnaissance study is usually a coarse screen intended to eliminate those projects or actions which are clearly infeasible without extensive study, and hence to identify those activities which deserve further study. Following the reconnaissance phase may be one or more phases intended to thoroughly evaluate the feasibility of the proposed activity and in the process to formulate a description of the most desirable actions, i.e. the plan. In many cases a single feasibility study is adequate because the nature of the action is relatively easily evaluated. In other cases one or more pre-feasibility studies may be undertaken to examine various aspects of the proposal. The idea of several sequential studies is to reduce planning costs by testing the weakest aspects of the project first. If the project is eliminated because of some aspect, the expense of studying all other aspects will have been avoided. If the series of studies is allowed to become a series of increasingly more thorough reviews of all aspects of the project, the cost may be increased because many things will have been redone two or more times.

The feasibility study usually requires that the structural details of a project be specified in sufficient detail to permit an accurate cost estimate. In the final phase of the effort, the details of design must be examined carefully, and the construction drawings and specifications produced. Although, in principle, the issue of feasibility was decided on the basis of the feasibility study, the possibility always exists that the more thorough study may develop information that alters feasibility. Consequently, the decision to proceed should not be made irrevocable until the final design is complete.

**(to be continued)**

## New Words and Expressions

1.inevitable *a.* 不可避免的，必然的，料得到的；-ly *ad.*

2.connotation *n.* 含蓄，涵义，内涵

3.obsolete *a.* 已抛弃（不用）的，过时的

4.periodical *a.* 定期的，周期的，期刊的；-ly *ad.*

5.coordinate *v.* （使）成为同等，（使）协调

6.enhance *vt.* 增加，提高；-ment *n.*

7.legislative *n.* 有立法权（的）

8.executive a. 执行的，行政的 ~branch 行政部门
9.authority n. 管理局（机构），上级，（pl.）当局，权力（限），代理权
10.isolation n. 孤立，单独，隔离；vt. isolate
11.legitimate a. 合法的，合理的，真实的；-meit vt. 使合法
12.emerge vi. 浮现，形成，发生，排出
13.reconnaissance n. 侦察，勘测，草测，选点，调查研究；~ study 设计任务书阶段
14.eliminate vt. 删去，淘汰，排除
15.screen n. 屏，筛分（选）
16.identify v. 识（鉴）别，认出，被等同
17.deserve vt. 应受（得），值得
18.formulate vt. 公式化，用式表示，配制，系统阐述，明确表示，扼要订出，正式提出，定义
19.description n. 叙述，说明（书）
20.expense n. 开销，花费，~at sb. 归某人负担；（pl.）经费
21.expend vt.同 spend（花钱时常用）
22.specify vt. 指定，详细说明
23.specification n. 详细说明，（pl.）规格，明细单，说明书
24.issue n.流出，发行（额），报刊期号，问题，结果，结局
25.irrevocable a.不能改变（挽回，取消），最后的
26.approach to （做某事，解决某问题）方法（途径）
27.master plan （地区）总体规划
28.most desirable pattern 最适宜的模式
29.not more than 至多，不多于
30.to allow for 考虑到，便于，为…创造条件
31.allow of 容许，可以有
32.feasibility study 可行性研究

## Reading Natural Planning for Water Resources Development

**(continue)**

To say that a water project is feasible implies that it will effectively serve its intended purpose without serious negative impact external to the project. Hence, to measure feasibility, the project purpose(s) or objective(s) must be specified before the planning begins. The rules for measuring achievement must also be specified. At the national level the objectives are usually rather broad. The U. S. national objectives for water planning are (1) enhancement of national economic development and (2) enhancement of the quality of the environment. Other objectives might be to increase national food production, encourage regional development, improve transportation, etc.

At the river basin or project level these objectives may be translated into more specific goals. For example, food production may be increased by irrigation, land drainage, and flood protection as well as by non-water-related actions such as provision of more fertilizer, education of farmers, improved seeds, etc. Planning should test all possible alternative measures or combination of these measures and the objectives should remain broad until these tests clearly eliminate some of the measures. Thus the objective should not be reduced to provision of irrigation until it is clear that other alternatives are not viable or at least little attractive.

Budgets for planning are usually limited, and before planning at the project level actually starts a strategy for the particular situation should be developed. To do this, the problems should be assessed as carefully as possible and the factors most likely to be critical in shaping the plan identified. For example, in an arid area the availability of water will be a critical factor in the feasibility of an irrigation project. Unless available data are clearly sufficient to evaluate water needs and sources, the hydrologic analysis may be the controlling aspect of planning. In other cases, the character of the soils, availability of reservoir sites, or even the adequacy of the foundation at a possible dam site may govern. The allocation of effort in the planning process should be such as to clearly answer those questions which control the decision. Detailed investigation of other topics may be left to a later stage. Selection of staff, programs of data assembly, and the scheduling of work should be planned in the light of the assessment of critical problems, so that when the funds have been spent some definitive answers will be available. If this is not done, subsequent studies will have to re-do much of the work of the first study and the final decision will have to be postponed and the overall cost of planning increased.

An important consideration in the strategy for planning is the need for data. Most of the data required for planning are current data describing existing conditions of land use, population, topography, etc. These data must be available when needed but can be collected at any time prior to this need. Hydrologic and climatological data are historic and must be collected over a time period prior to their use. Consequently, if the available historic data appear inadequate, installation of new stations should be undertaken at the beginning of planning in order that some data will be available. Assembly of less urgently needed data should be postponed if necessary.

Since planning is always for the future, forecasts of future conditions are essential. Since the planning horizon for water projects is usually 50 to 100 yr, forecasts are particularly critical. Unfortunately, no forecast can be perfect and reflection on the changes in the past 50 yr.

It is not likely that significant improvement in forecast methods will occur. Even though more rigorous procedures are used, the results will still be dependent on assumed constancy of rate coefficients which have never remained constant. Simple trend extrapolations should be avoided in favor of more rational forecast methods, wherever possible, but the planning problem is that of minimizing the risk of a wrong decision as the result of a poor forecast. This can be done by considering a range of forecast values or alternative futures. The range may be specified by varying factors in the forecast process, or by preparing scenarios for the future and evaluating the forecast in the light of these scenarios. The various project alternatives are tested to see how they perform under the range of future conditions. An alternative which is robust enough to perform well regardless of the future assumed is likely to be the best alternative. If no alternative is satisfactory, whatever the outcome, then it will be necessary to reformulate an alternative which is flexible. This might be done by stage construction, i. e, building a small dam with the possibility of raising it at a later date, or installing only a portion of the turbines at a power plant. The use of groundwater for urban water supply or irrigation may be a useful, low-cost, first-stage alternative. If subsequent growth of demand justifies a reservoir, well and good, if not, no investment is lost.

Possible changes in technology should always be considered in projections of the future. Obviously not all changes which might occur over the next 50 or 100 yr will be anticipated, but shorter term changes should be evident. It is not good planning to suggest a project which will be technologically obsolete before it is completed. In some cases, a delay may permit later construction of a far superior project.

Once the basic data and the projections of future conditions are assembled, actual formulation of the project can commence. This is a phase of planning where imagination and skill are required. The important first consideration is the compilation of a list of alternatives. Such a list should be comprehensive. After a preliminary project is in hand as the first phase of the planning operation, and the operation is reduced to attempting to justify this plan. This approach is wrong. The planning process should be an evaluation of all possible alternatives with respect to project features and water use. It is not even sufficient to start with the assumption that a specified quantity of water is required and to evaluate the merits of two or more plans for supplying this quantity. Planning should consider alternative and competing uses for water as well as the various possibilities for control and delivery of the water. The planning process may involve considerable feedback. As project formulation proceeds, it may be evident that new data or projections are required or that some revision of background data is needed.

The first step in project formulation is the definition of the boundary conditions which restrict the project. For example:

(1) One or more aspects of water development can be eliminated on the basis of physical limitations, i. e., no navigation on torrential mountain streams.

(2) Certain problems may be fixed in location, i. e., flood mitigation for an existing city.

(3) The available water may be limited or subject only to minor changes.

(4) Maximum land areas usable for various purposes may be definable. This does not exclude possibility of alternative uses for a given parcel of land.

(5) A policy decision may reserve certain lands for specific purposes, i. c., parks and recreation areas.

(6) Possible sites for water storage (both surface and underground) can be defined and their limiting capacities evaluated.

(7) Certain existing foci of water use exist and must continue to be supplied.

(8) Legal constraints may reserve certain lands or prohibit certain activities or actions.

In the United States, planning agencies are required to seek the participation of interested members of the public in the planning process. Suggestions from the public may point to possible alternatives not considered by the planners, and the views of the public are especially useful in the evaluation. The role of planners is to present alternatives for consideration of the public or their elected decision makers. Planners must be careful not to eliminate an alternative because of their own views or prejudices.

## New Words and Expressions

1.provision *n.* 预备，保证，供给，条款

2.assess *vt.* 估价（财产），确定，查定；-ment *n.*

3.assembly *n.* 集合，装配，议会

4.postpone *vt.* 延期，搁置到（until, to），延期（for），放在次位（to）

5.urgent *a.* 紧急（迫）的；-ly *ad.*

6.be in urgent need 急需

7.rigorous *a.* 严格的，精密的

8.constancy *n.* 不变性，固定性

constant *a.*；*n.* 常数（值）

9.extrapolate *v.* 推断，外推；-tion *n.*

10.scenario *n.* 剧情说明，剧本，方案

11.robust *a.* 强健的，茁壮的，耐用的

12.flexible *a.* 易适应的，灵活的

13.anticipate *vt.* 预期，在…之先；-tion *n.*

14.evidence *n.*; *vt.* 明显，根据，证据

in- 显著的

15.evident *a.* 明显的，明白的； -ly *ad.*

16.commence *v.* 开始，倡导； -ment *n.*

17.imagination *n.*想像(力)，创造力，空想 imagine *v.*

18.compile *vt.* 编辑，搜集，汇编 -tion *n.*

19.merit *n.* 优点，特征，指标，价值 （*pl.*）功过，是非；*vt.* 有…价值

20.feedback *n.* 反馈，反应

21.torrential *a.* 急流的

22.navigate *v.* 航行（于），驾驶；-tion *n.*

23.mitigate *vt.* 使缓和，减轻，调节（冷热）；-tion *n.*

24.recreate *v.* （使） 休养，娱乐；-tion *n.*

25.focus *n.* 焦点，焦距，中心；*v.*

26.versus *prep.* 对，依…为转移，作为…的函数

27.prejudice *prep.*；*n.*偏见，歧视；*vt.*

# Lesson 8 Reservoirs

When a barrier is constructed across some river in the form of a dam, water gets stored up on the upstream side of the barrier, forming a pool of water, generally called a reservoir.

Broadly speaking, any water collected in a pool or a lake may be termed as a reservoir. The water stored in reservoir may be used for various purposes. Depending upon the purposes served, the reservoirs may be classified as follows: ① Storage or Conservation Reservoirs; ② Flood Control Reservoirs; ③ Distribution Reservoirs; ④ Multipurpose Reservoirs.

## 1 Storage or Conservation Reservoirs

A city water supply, irrigation water supply or a hydroelectric project drawing water directly from a river or a stream may fail to satisfy the consumers demands during extremely low flows, while during high flows, it may become difficult to carry out their operation due to devastating floods. A storage or a conservation reservoir can retain such excess supplies during periods of peak flows and can release them gradually during low flows as and when the need arises.

Incidentally, in addition to conserving water for later use, the storage of floodwaters may also reduce flood damage below the reservoir. Hence, a reservoir can be used for controlling floods either solely or in addition to other purposes. In the former case, it is known as "Flood Control Reservoir" or "Single Purpose Flood Control Reservoir", and in the later case, it is called a "Multipurpose Reservoir".

## 2 Flood Control Reservoirs

A flood control reservoir or generally called flood-mitigation reservoir, stores a portion of the flood flows in such a way as to minimize the flood peaks at the areas to be protected downstream. To accomplish this, the entire inflow entering the reservoir is discharged till the outflow reaches the safe capacity of the channel downstream. The inflow in excess of this rate is stored in the reservoir, which is then gradually released so as to recover the storage capacity for next flood.

The flood peaks at the points just downstream of the reservoir are thus reduced by an amount AB as shown in Fig.8.1. A flood control reservoir differs from a conservation reservoir only in its need for a large sluiceway capacity to permit rapid drawdown before or after a flood.

Types of flood control reservoirs. There are two basic types of flood-mitigation reservoirs. ①Storage Reservoirs or Detention basins.②Retarding basins or retarding reservoirs.

A reservoir with gates and valves installation at the spillway and at the sluice outlets is known as a storage-reservoir, while on the other hand, a reservoir with fixed ungated outlets is known as a retarding basin.

Functioning and advantages of a retarding basin:

A retarding basin is usually provided with an uncontrolled spillway and an uncontrolled orifice type sluiceway. The automatic regulation of outflow depending upon the availability of water, takes place from such a reservoir. The maximum discharging capacity of such a reservoir should be equal to the maximum safe carrying capacity of the channel downstream. As flood occurs, the reservoir gets filled and discharges through sluiceways. As the reservoir elevation increases, outflow discharge increases. The water level goes on rising until the flood has subsided and the inflow becomes equal to or less than the outflow. After this, water gets automatically withdrawn from the reservoir until the stored water is completely discharged. The advantages of a retarding basin over a gate controlled detention basin are: ①Cost of gate installations is saved. ②There are no gates and hence, the possibility of human error and negligence in their operation is eliminated. ③Since such a reservoir is not always filled, much of land below the maximum reservoir level will be submerged only temporarily and occasionally and can be successfully used for agriculture, although no permanent habitation can be allowed on this land.

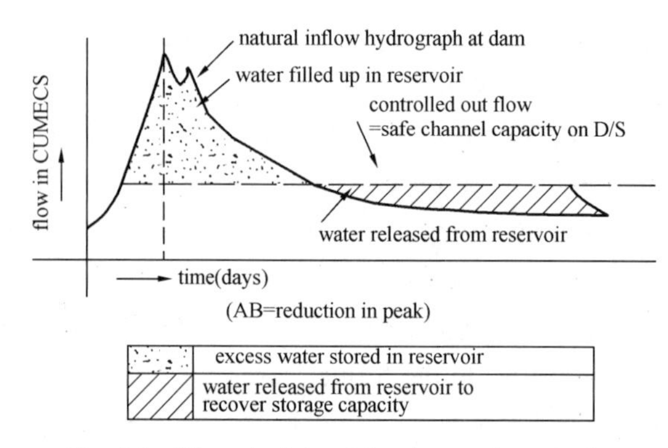


Fig. 8.1 The principle of flood control reservoirs

Functioning and advantages of a storage reservoir:

A storage reservoir with gated spillway and gated sluiceway, provides more flexibility of operation, and thus gives us better control and increased usefulness of the reservoir. Storage reservoirs are, therefore, preferred on large rivers which require better control, while retarding basins are preferred on small rivers. In storage reservoirs, the flood crest downstream can be better controlled and regulated properly so as not to cause their coincidence. This is the biggest advantage of such a reservoir and outweighs its disadvantages of being costly and involving risk of human error in installation and operation of gates.

### 3 Distribution Reservoirs

A distribution reservoir is a small storage reservoir constructed within a city water supply system. Such a reservoir can be filled by pumping water at a certain rate and can be used to supply water even at rates higher than the inflow rate during periods of maximum demands (called critical periods of demand). Such reservoirs are, therefore, helpful in permitting the pumps or water treatment plants to work at a uniform rate, and they store water during the hours of no demand or less demand and supply water from their "storage" during the critical periods of maximum demand.

### 4 Multipurpose Reservoirs

A reservoir planned and constructed to serve not only one purpose but various purposes together is called a multipurpose reservoir. Reservoir, designed for one purpose, incidentally serving other purposes, shall not be called a multipurpose reservoir, but will be called so, only if designed to serve those purposes also in addition to its main purpose. Hence, a reservoir designed to protect the downstream areas from floods and also to conserve water for water supply, irrigation, industrial needs, hydroelectric purposes, etc. Shall be called a multipurpose reservoir.

## New Words and Expressions

1.upstream *a.*; *ad.* （向，在）上游（的）
2.devastate *vt.* 使…荒废
3.solely *ad.* 惟一地，单独地
4.inflow *n.* 进水量；流入，入流
5.outflow *n.* 流出量；放水，出流
6.recover *vt.* 恢复
7. sluiceway *n.*泄洪道
8.valve *n.* 阀门，开关
9.ungated *a.* 无门的；闭塞的
10.orifice *n.* 孔口；管口
11.negligence *n.* 疏忽；粗心大意
12.eliminate *vt.* 消（排，清）除
13.temporarily *ad.* 暂（临）时地
14.flexibility *n.* 机动性；灵活性
15.coincidence *n.* 符合；一致；同时发生
16.outweight *vt.* 重于，比…（重要）
17.costly *a.* 昂贵的，重价的
18.incidentally *ad.* 伴随地，附带地
19.broadly speaking 概括地说
20.storage reservoir 蓄水库
21.flood control reservoir 防洪水库
22.distribution reservoir 配水水库
23.multipurpose reservoir 综合利用水库
24.devastating floods 毁灭性洪水
25.peak flow 洪峰流量，洪水
26.flood-mitigation reservoir 减洪水库，滞洪水库
27.flood peak 洪峰
28.safe capacity of channel downstream 下游河道安全泄量
29.storage capacity 蓄水库容，库容
30.detention basin 蓄洪区；拦洪区
31.retarding basin 滞洪水库
32.flood crest 洪峰
33.critical period of demand 需水临界期

## Reading Material The Grand Three Gorges Project

**Three Gorges** project is mainly composed of a river-barrage dam, hydropower station and navigation structures.

### 1 River-barrage Dam

The dam is a concrete gravity dam, pulling up from the river base to the dam crest at an elevation of 185m, the maximum dam height from its foundation to the crest is 175m. The total length of dam axis is 2335m, cutting the river flow laterally to form a lofty "great-wall on river", damming a head of 80m (flood season) to 110m (dry season). The elevation of maximum storage level is 175m. After the dam is in use, a deep navigation channel with a length of 600m and a river-channel-type reservoir will be formed, the water will directly trace back to Chongqing and turn this hilly city into a fine inland harbor.

The dam can be divided into several sections. Locating at the middle section, i. e., the original position of the main river trough, is the flood-relieving dam section of 483m in length, equipped with discharging sluices having totally twenty-three 7m×9m flood-relieving deep holes and twenty-two flood-relieving surface holes of 8m net width. Each hole has a control gate to be opened or closed freely, which is able to discharge flow under various requirements. The maximum flood releasing capability is designed at 100,000$m^3$/s which is equivalent to that of Gezhouba. This capacity is even met the historical extraordinary flood happened in 1870 and therefore, is able to discharge flood safely and ensure the safety of the dam.

## 2 Hydropower Station

The hydropower station at the toe of the dam is situated respectively at both sides of the flood-relieving dam section. The installation capacity of a unit is 700,000 kW. The power station has 26 units with a total installation capacity of 18.2 million kW. The total length of the plant building is 643m at leftside with 14 generating units installed and is 576m at right-side with 12 generating units installed. An average annual generating capacity of the power station is 84.7 billion kW·h. The water-control dam section at both sides of the power station is connected to the hills at two banks. To consider the expansion requirement of the installation capacity, spaces of the underground plant building are reserved under the dam, which is able to install 6 units of 700,000 kW generators.

## 3 Navigation Structures

Two permanent navigation structures are arranged at the left bank. One is the double channel 5-stage ship lock. The long-term plan of one-way passing ability is 51.5 million tons per year. The other is the vertical-lifting ship-elevator which is able to pass a passenger-cargo vessel of 3,000 ton-class.

On the deck between the dam and the plant building, the main transformer has been installed to boost the voltage to 500,000 volts and is stretched to the tower of lower reaches at left and right banks. The transformer is the outlet point of the power station and 15 extra-high voltage transmission lines starting from here to transmit powerful electric current to mid-China, east-China and east-Sichuan areas. After the completion of the construction of Three Gorges hydropower station, it will serve as an important supporting point to create significant factors for forming a huge combined network of whole nation's electric networks.

# New Words and Expressions

1.project *n.* 计划，规划，设计，工程，项目

2.dam *n.* （水，挡水）坝，障碍，挡在坝里的水
*vt.*（dammed; damming）（挡）起，筑坝（拦水）

3.navigation *n.* 导（领）航，航行（学），船舶（总称）

4.concrete *n.* 混凝土，三合土，凝结物，具体物

5.gravity *n.* 重力，万有引力

6.base *n.* 基（底，础，本，质，托，板，体，准，点，线，面，座，地）

7.crest *n.* 顶，顶峰，振幅，峰值

8.elevation *n.* 高强，标高，海拔

9.laterally *ad.* 在[由]侧面，横向地

10.lofty *a.* 高（耸）的，极高的，崇高的，高级的，玄虚的

11.channel *n.* 沟道，渠道，管道，沟渠，河床，沟槽，河槽

12.harbor *n.* 海港，港湾[口]，码头

13.inland *n.*; *a.* 内地[陆]（的），国内（的）；在[到，向]内地

14.section *n.* 截（断，剖，切）面，分段

15.shiplock *n.* 船闸

16.vessel *n.* 容[贮]器，槽，船[只]，运输机

17.deck *n.* 甲[舱，面]板，舱面，台面，（平）台

18.transformer *n.* 变压器，互感器

19.boost *vt.*; *n.* 推起，升高，放大，升压

20.voltage *n.* 电压（量），电位（势）差，伏（特）数

21.stretch *n.* 伸（展，长，出，张，延），扩大，张开

22.dam up 壅高，用坝堵高水位

23.dam out 筑坝排水

24.navigation lock 船闸

# Lesson 9 Properties of Concrete

The characteristics of concrete should be considered in relation to the quality required for any given construction purpose. The closest practicable approach to perfection in every property of the concrete would result in poor economy under many conditions, and the most desirable structure is that in which the concrete has been designed with the correct emphasis on each of the various properties of the concrete, and not solely with a view to obtaining, say, the maximum possible strength.

Although the attainment of the maximum strength should not be the sole criterion in design, the measurement of the crushing strength of concrete cubes or cylinders provides a means of maintaining a uniform standard of quality, and, in fact, is the usual way of doing so. Since the other properties of any particular mix of concrete are related to the crushing strength in some manner, it is possible that as a single control test it is still the most convenient and informative.

The testing of the hardened concrete in prefabricated units presents no difficulty, since complete units can be selected and broken if necessary in the process of testing. Samples can be taken from some parts of a finished structure by cutting cores, but at considerable cost and with a possible weakening of the structure. It is customary, therefore, to estimate the properties of the concrete in the structure on the basis of the tests made on specimens moulded from the fresh concrete as it is placed. These specimens are compacted and cured in a standard manner given in BS 1881: 1970 as in these two respects it is impossible to simulate exactly the conditions in the structure. Since the crushing strength is also affected by the size and shape of a specimen or part of a structure, it follows that the crushing strength of a cube is not necessarily the same as that of the mass of exactly the same concrete.

## 1 Crushing strength

Concrete can be made having a strength in compression of up to about 80N/mm$^2$ (12,000lb/in$^2$), or even more depending mainly on the relative proportions of water and cement, that is, the water/cement ratio, and the degree of compaction. Crushing strengths of between 20 and 50N/mm$^2$ at 28 days are normally obtained on the site with reasonably good supervision, for mixes roughly equivalent to 1:2:4 of cement:sand:coarse aggregate. In some types of precast concrete such as railway sleepers, strengths ranging from 40 to 65 N/mm$^2$ at 28 days are obtained with rich mixes having a low water/cement ratio.

The crushing strength of concrete is influenced by a number of factors in addition to the water/cement ratio and the degree of compaction. The more important factors are:

**(1) Type of cement and its quality.** Both the rate of strength gain and the ultimate strength may be affected.

**(2) Type and surface texture of aggregate.** There is considerable evidence to suggest that some aggregates produce concrete of greater compressive and tensile strengths than are obtained with smooth river gravels.

**(3) Efficiency of curing.** A loss in strength of up to about 40 per cent may result from premature drying out. Curing is therefore of considerable importance both in the field and in the making of tests. The method of curing concrete test cubes given in BS 1881 should, for this reason, be strictly adhered to.

**(4) Temperature.** In general, the rate of hardening of concrete is increased by an increase in temperature. At freezing temperatures the crushing strength may remain low for some time.

**(5) Age.** Under normal conditions concrete increases in strength with age, the rate of increase depending on the type of cement. For instance, high alumina cement produces concrete with a crushing strength at 24 hours equal to that of normal portland cement concrete at 28 days. Hardening continues but at a much slower rate for a number of years.

The above refers to the static ultimate load. When subjected to repeated loads concrete fails at a load smaller than the ultimate static load, a fatigue effect. A number of investigators have established that after several million cycles of loading, the fatigue strength in compression is 50-60 per cent of the ultimate static strength.

## 2 Tensile and flexural strength

The tensile strength of concrete varies from one-eighth of the compressive strength at early ages to about one-twentieth later, and is not usually taken into account in the design of reinforced concrete structures. The tensile strength is, however, of considerable importance in resisting cracking due to changes in moisture content or temperature. Tensile strength tests are used for concrete roads and airfields.

The measurement of the strength of concrete in direct tension is difficult and is rarely attempted. Two more practical methods of assessing tensile strength are available. One gives a measure of the tensile strength in bending, usually termed the flexural strength. BS 1881:1970 gives details concerning the making and curing of flexure test specimens, and of the method of test. The standard size of specimen is 150mm×150mm×750mm long for aggregate of maximum size 40mm. If the largest nominal size of the aggregate is 20mm, specimens 100mm×100mm×500mm long may be used.

A load is applied through two rollers at the third point of the span until the specimen breaks. The extreme fiber stresses, that is, compressive at the top and tensile at the bottom, can then be computed by the usual beam formulae. The beam will obviously fail in tension since the tensile strength is much lower than the compressive strength. Formulae for the calculation of the modulus of rupture are given in BS 1881:1970.

Test specimens in the form of beams are sometimes used to measure the modulus of rupture or flexural strength quickly on the site. The two halves of the specimen may then be crushed so that besides the flexural strength the compressive strength can be approximately determined on the same sample. The test is described in BS 1881:1970.

Values of the modulus of rupture are utilized in some methods of design of unreinforced concrete roads and runways, in which reliance is placed on the flexural strength of the concrete to distribute concentrated loads over a wide area.

More recently introduced is a test made by splitting cylinders by compression across the diameter to give what is termed the splitting. tensile strength. Details of the method are given in BS 1881:1970. The testing machine is fitted with an extra bearing bar to distribute the load along the full length of the cylinder Plywood strips, 12mm wide and 3mm thick are inserted between the cylinder and the testing machine bearing surfaces top and bottom. From the maximum applied load at failure the tensile splitting strength is calculated as follows:

$$f_t = \frac{2P}{\pi l d}$$

where: $f_t$ is splitting tensile strength, N/mm$^2$; $P$ is maximum applied load, N; $l$ is length of cylinder, mm; $d$ is diameter, mm.

As in the case of the compressive strength, repeated loading reduces the ultimate strength so that the fatigue strength in flexure is 50-60 per cent of the static strength.

### 3 Shear strength

In practice, shearing of concrete is always accompanied by compression and tension caused by bending, and even in testing it is impossible to eliminate an element of bending.

## New Words and Expressions

1.perfection *n.* 完善，完整性
2.sole *a.* 惟一的
3.criterion *n.* 标准，依据，准则
4.crushing *n.* 挤压，压碎
5.cylinder *n.* 圆柱体；圆筒
6.uniform *a.* 均质的，均匀的
7.mix *n.* 配合比
8.informative *a.* 有益的
9.prefabricate *vt.* 预制
10.unit *n.* 构件
11.customary *a.* 通常的，习惯的
12.mould *vt.* 浇筑
13.compact *vt.* 浇筑；压缩；使紧密
14.cure *vt.*（混凝土）养护
15.simulate *vt.* 模拟；模仿
16.supervision *n.* 监督；管理
17.aggregate *n.* 骨料
18.precast *vt.* 预制
19.sleeper *n.* 枕木，轨枕
20.texture *n.* 结构
21.tensile *a.* 张力的，拉力的
22.premature *a.* 不到时间的
23.adhere *v.* 遵守，粘着（于）
24.alumina *n.* 矾土，铝土
25.site-made *a.* 现场制成的
26.assess *vt.* 估值
27.flexural *a.* 弯（挠）曲的
28.flexure *n.* 弯（挠，屈，扭）曲
29.span *n.* 跨度；间隔
30.modulus *n.* 模量，模数
31.plywood *n.* 胶合板
32.strip *n.* 薄片
33.shearing *n.* 剪（断）切
34.with the emphasis on 着重，强调
35.with a view to 目的在于，为了
36.crushing strength 破坏强度
37.It is customary to (do…) 一般习惯于（做）…
38.It follows that… 由此得出…；因此，从而

39.water/cement ratio 水灰比
40.degree of compaction 压实度，密实度
41.equivalent to 相当于
42.adhere to 遵守
43.static ultimate loads = ultimate static load 极限静荷载
44.flexural strength 抗弯强度
45.reinforced concrete 钢筋混凝土
46.direct tension 轴向受位
47.fibre stress 纤维应力；边缘应力
48.compressive strength 抗压强度
49.modulus of rupture 断裂模量
50.splitting tensile strength 劈裂抗拉强度
51.BS=British Standards 英国标准
（=Bureau of Standards 标准局）
52.Portland cement 波特兰水泥
53.$N/mm^2$ 牛顿/毫米 $^2$
54.$lb/in^2$=$pound/inch^2$ 磅/英寸 $^2$

# Reading Material Durability of Concrete

Besides its ability to sustain loads, concrete is also required to be durable. The durability of concrete can be defined as its resistance to deterioration resulting from external and internal causes. The external causes include the effects of environmental and service conditions to which concrete is subjected, such as weathering, chemical actions and wear. The internal causes are the effects of salts, particularly chlorides and sulphates, in the constituent materials, interaction between the constituent materials, such as alkali-aggregate reaction, volume changes, absorption and permeability.

In order to produce durable concrete care should be taken to select suitable constituent materials. It is also important that the mix contains adequate quantities of materials in proportions suitable for producing a homogeneous and fully compacted concrete mass.

## 1 Weathering

Deterioration of concrete by weathering is usually brought about by the disruptive action of alternate freezing and thawing of tree water within the concrete and expansion and contraction of the concrete, under restraint, resulting from variations in temperature and alternate wetting and drying.

Damage to concrete from freezing and thawing arises from the expansion of pore water during freezing; in a condition of restraint, if repeated a sufficient number of times, this results in the development of hydraulic pressure capable of disrupting concrete. Road kerbs and slabs, dams and reservoirs are very susceptible to frost auction.

The resistance of concrete to freezing and thawing can be improved by increasing its impermeability. This can be achieved by using a mix with the lowest possible water-cement ratio compatible with sufficient workability for placing and compacting into a homogeneous mass. Durability can be further improved by using air entrainment, an air content of 3 to 6 per cent of the volume of concrete normally being adequate for most applications. The use of air-entrained concrete is particularly useful for roads where salts are used for deicing.

## 2 Chemical Attack

In general, concrete has a low resistance to chemical attack. There are several chemical agents which react with concrete but the most common forms of attack are those associated with leaching, carbonation, chlorides and sulphates. Chemical agents essentially react with certain compounds of the hardened cement paste and the resistance of concrete to chemical attack therefore can be affected by the type of cement used. The resistance to chemical attack improves with increased impermeability.

## 3 Wear

The main causes of wear of concrete are the cavitation effects of fast moving water, abrasive material in water, wind blasting and attrition and impact of traffic. Certain conditions of hydraulic flow result in the formation of cavities between the flowing water and the concrete surface. These cavities are usually filled with water vapor charged with extraordinarily high energy and repeated contact with the concrete surface results in the formation of pits and holes, known as cavitation erosion. Since even a good-quality concrete will not be able to resist this kind of deterioration, the best remedy is therefore the elimination of cavitation by producing smooth hydraulic flow. Where necessary, the critical areas may be lined with materials having greater resistance to cavitation erosion.

In general, the resistance of concrete to erosion and abrasion increases with increase in strength. The use of a hard and tough aggregate tends to improve concrete resistance to wear.

## 4 Alkali-Aggregate Reactions

Certain natural aggregates react chemically with the alkalis present in Portland cement. When this happens, these aggregates expand or swell resulting in cracking and disintegration of concrete.

## 5 Volume Changes

Principal factors responsible for volume changes are the chemical combination of water and cement and the subsequent drying of concrete, variations in temperature and alternate wetting and drying. When a change in volume is resisted by internal or external forces, this can produce cracking, the greater the imposed restraint, the more severe the cracking. The presence of cracks in concrete reduces its resistance to the action of leaching, corrosion of reinforcement, attack by sulphates and other chemicals, alkali-aggregate reaction and freezing and thawing, all of which may lead to disruption of concrete. Severe cracking can lead to complete disintegration of the concrete surface particularly when this is accompanied by alternate expansion and contraction.

Volume changes can be minimized by using suitable constituent materials and mix proportions due to the size of structure. Adequate moist curing is also essential to minimize the effects of any volume changes.

## 6 Permeability and Absorption

Permeability refers to the ease with which water can pass through the concrete. This should not be confused with the absorption property of concrete and the two are not necessarily related. Absorption may be defined as the ability of concrete to draw water into its voids. Low permeability is an important requirement for hydraulic structures and in some cases water-tightness of concrete may be considered to be more significant than strength although, other conditions being equal, concrete of low permeability will also be strong and durable. Concrete which readily absorbs water is susceptible to deterioration.

Concrete is inherently a porous material. This arises from the use of water in excess of that required for the purpose of hydration in order to make the mix sufficiently workable and the difficulty of completely removing all the air from the concrete during compaction. If the voids are interconnected concrete becomes pervious although with normal care concrete is sufficiently impermeable for most purposes. Concrete of low permeability can be obtained by suitable selection of its constituent materials and their proportions followed by careful placing, compaction and curing. In general for fully compacted concrete, the permeability decreases with decreasing water-cement ratio. Permeability is affected by both the fineness and the chemical composition of cement. Coarse cements tend to produce pastes with relatively high porosity. Aggregates of low porosity are preferable when concrete with a low permeability is required. Segregation of the constituent materials during placing can adversely affect the impermeability of concrete.

## New Words and Expressions

1.durable *a.* 耐久的，经久的，永久性的；*n.(pl.)* 耐用的物品
2.durability *n.* 耐久性，经久性，耐用年限
3.chloride *n.* 氯化物，漂白粉
4.sulphate *n.* 硫酸盐
5.alkali *n.* 碱，碱性，强碱
6.absorption *n.* 吸收，吸取，吸水性
7.permeability *n.* 渗透性，透气性，渗透
8.deterioration *n.* 变坏，变质，损坏，损伤
9.weathering *n.* 风化（作用，层），自然老化，大气侵蚀
10.disruptive *a.* 分裂的，摧毁的，破坏的
11.thaw *v.* 使融化，解冻，溶化
12.entrain *v.* 携带，传输，使（空气）以气泡状存于混凝土中，产生
13.deicing *n.* 除冰，防冻（装置）
14.leaching *n.* 浸出，浸析作用，浸滤，溶析
15.carbonation *n.* 碳化作用，碳酸盐化
16.blasting *n.* 破裂，吹风，气流加速运动，喷砂
17.attrition *n.* 磨损，磨耗，损耗
18.hydraulic *a.* 水力的，液力的，水压的；*n.* 水力
19.cavity *n.* 空腔，空穴，孔穴，洞穴
20.cavitation *n.* 气蚀，空蚀，空化作用
21.impermeability *n.* 不渗透性，防水性，气密性
22.hydraulic structure 水工建筑物
23.pervious *a.* 透水的，透光的，有孔的，能通过的
24.aggregate *n.* 集料，骨料，总计，合计
25.homogeneous *a.* 均质的，均匀的，单相的
26.compact *a.* 紧密的，密实的；*v.* 压实，捣实
27.kerb *n.* 路缘，道牙，建筑物上的边饰
28.air-entrained concrete 加气混凝土

# Lesson 10 Basic Concepts of Reinforced Concrete

Although concrete is used very extensively in the construction of buildings, bridges and many other engineering structures, its mechanical properties are far from ideal. For example, it is not a particularly strong material. The compressive strength of structural grade concrete ranges typically from 20 to 40 MPa, or about 3000 to 6000 lbs/in$^2$. This is somewhat lower than the compressive strength range of most timbers used in structural work. The tensile strength of concrete is extremely low, about one-tenth of its compressive strength, and this precludes the use of plain concrete for most structural members. The elastic modulus for concrete subjected to compressive stress of short duration is reasonably high, in the range of 20,000 to 30,000 MPa (about one-tenth of the elastic modulus of steel); however, concrete undergoes large additional long-term deformations due to creep and shrinkage, so that the effective stiffness is much lower—perhaps a third to a quarter of the instantaneous stiffness.

The widespread use of concrete in engineering construction stems from its cheapness compared with other structural materials currently available. Its lack of tensile strength is overcome by including reinforcement, usually in the form of steel bars, to produce a composite material known as reinforced concrete. Although the steel reinforcement does not prevent cracking of the concrete in regions of tension, it does prevent the cracks from widening, and it provides an effective means for resisting the internal tensile forces. The quantity of reinforcement need is usually quite small, relative to the volume of concrete, so that the total cost of reinforced concrete construction remains commercially very competitive.

The primary purpose of the steel reinforcement is to carry internal tensile forces. Reinforcement is therefore placed in beams near the tensile face, i. e, near the lower face in the in-span regions of positive moment and near the upper face in regions near internal supports, where negative moments act. This is illustrated in Fig. 10.1. In reinforced concrete design, it is important to provide reinforcement in all regions of potential cracking. Thus a rectangular arrangement of vertical and horizontal steel bars is introduced into regions of a beam where inclined cracks can form as a result of combined shearing action and bending moment. The longitudinal steel, or main reinforcement, and the transverse bars, called stirrups, may be preassembled into a reinforcing cage for ease of construction.

In a floor slab, bars are usually laid in the two main span directions at right angles, to resist the tensile forces produced by the bending actions in each direction. For ease of construction, welded mesh is frequently used as slab reinforcement.

Although steel reinforcement is used primarily to carry the internal tensile forces produced by external loading, it also has other uses. Steel is much stronger than concrete in compression, and it is sometimes used to boost the resistance of zones of compression, when the overall dimensions of the member are restricted. Longitudinal steel is placed in all compression members. In such members, bending is usually present in addition to axial compression and the longitudinal steel reinforcement at each face of the member may act either in tension or compression. Transverse ties are used to maintain the longitudinal steel in position during casting of the concrete and later to prevent its outward buckling when it is subjected to compressive stress. Again, the reinforcing steel for the column may be preassembled into a cage.

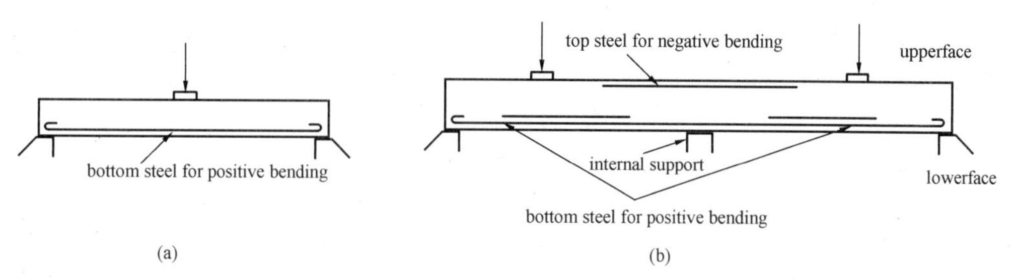


Fig.10.1 Main longitudinal reinforcement in beams

(a) Simply supported beam; (b) Continuous beam

Cracking in concrete may be caused not only by external loading, but also by temperature gradients and differential or restrained shrinkage. Secondary reinforcement is therefore provided to control such cracking, which may be unsightly and even dangerous.

## New Words and Expressions

1.preclude *vt.* 阻碍
2.restrain *vt.* 抑制
3.duration *a.* 短期的
4.instantaneous *a.* 瞬时的
5.commercial *a.* 商业上的
6.preassemble *v.* 预装
7.cage *n.* 骨架
8.boost *vt., n.* 提高；加强
9.buckling *n., a.* 屈曲
10.for ease in 为了便于；以便
11.lay in 搁置
12.at right angles 成直角

## Reading Material Reinforced Concrete

Plain concrete is formed from a hardened mixture of cement, water, fine aggregate, coarse aggregate (crushed stone or gravel), air, and often other admixtures. The plastic mix is placed and consolidated in the form-work, then cured to facilitate the acceleration of the chemical hydration reaction of the cement/water mix, resulting in hardened concrete. The finished product has high compressive strength, and low resistance to tension, such that its tensile strength is approximately one-tenth of its compressive strength. Consequently, tensile and shear reinforcement in the tensile regions of sections has to be provided to compensate for the weak-tension regions in the reinforced concrete element.

It is this deviation in the composition of a reinforced concrete section from the homogeneity of standard wood or steel sections that requires a modified approach to the basic principles of structural design. The two components of the heterogeneous reinforced concrete section are to be so arranged and proportioned that optimal use is made of the materials involved. This is possible because concrete can easily be given any desired shape by placing and compacting the wet mixture of the constituent ingredients into suitable forms in which the plastic mass hardens. If the various ingredients are properly proportioned, the finished product becomes strong, durable, and, in combination with the reinforcing bars, adaptable for use as main members of any structural system.

The techniques necessary for placing concrete depend on the type of member to be cast: that is, whether it is a column, a beam, a wall, a slab, a foundation, a mass concrete dam, or an extension of previously placed and hardened concrete. For beams, columns, and walls, the forms should be well oiled after cleaning them, and the reinforcement should be cleared of rust and other harmful materials. In foundations, the earth should be compacted and thoroughly moistened to about 6 in. in depth to avoid absorption of the moisture present in the wet concrete. Concrete should always be placed in horizontal layers, which are compacted by means of high-frequency power-driven vibrators of either the immersion or external type, as the case requires, unless it is placed by pumping. It must be kept in mind, however, that over-vibration can be harmful since it could cause segregation of the aggregate and bleeding of the concrete.

Hydration of the cement takes place in the presence of moisture at temperatures above 50℉. It is necessary to maintain such a condition in order that the chemical hydration reaction can take place. If drying is too rapid surface cracking takes place. This would result in reduction of concrete strength due to cracking as well as the failure to attain full chemical hydration.

It is clear that a large number of parameters have to be dealt with in proportioning a reinforced concrete element, such as geometrical width, depth, area of reinforcement, steel strain, concrete strain, steel stress, and so on. Consequently, trial and adjustment is necessary in the choice of concrete sections, with assumptions based on conditions at site, availability of the constituent materials, particular demands of the owners, architectural and headroom requirements, the applicable codes, and environmental conditions. Such an array of parameters has to be considered because of the fact that reinforced concrete is often a site-constructed composite, in contrast to the standard mill-fabricated beam and column sections in steel structures.

A trial section has to be chosen for each critical location in a structural system. The trial section has to be analyzed to determine if its nominal resisting strength is adequate to carry the applied factored load. Since more than one trial is often necessary to arrive at the required section, the first design input step generates into a series of trial-and-adjustment analyses.

The trial-and-adjustment procedures for the choice of a concrete section lead to the convergence of analysis and design. Hence every design is an analysis once a trial section is chosen. The availability of handbooks, charts, and personal computers and programs supports this approach as a more efficient, compact, and speedy instructional method compared with the traditional approach of treating the analysis of reinforced concrete separately from pure design.

# New Words and Expressions

1.harden *v.* （使）变硢[硬化，凝固，坚固]

2.aggregate *n.* 集料，骨料，集合

3.gravel *n.* 砾石，卵石

4.admixture *n.* 混合物，外加剂

5.consolidate *v.* 巩固，压实，捣实

6.cure *n.*，*v.* 治疗，处理，（混凝土）养护

7.hydration *n.* 水化作用

8.reinforcement *n.* 增强，补强，钢筋，配筋

9.homogeneity *n.* 同种，均匀性，一致性

10.deviation *n.* 偏差，偏异，差异

11.heterogeneous *a.* 不均匀的，非均质的

12.constituent *a.* 组织的；*n.* 组成，构成元素，组成物

13.ingredient *n.*（混合物）的成份，组成部分，配料

14.immersion *n.* 浸入，插入

15.segregation *n.* 分离，隔离，分开

16.parameter *n.* 参数，参量

17.assumption *n.* 假定，前提，承担

18.headroom *n.* 净空高度，头上空间

19.nominal *a.* 铭牌的，名义的

20.convergence *n.* 收敛，集中，会合，趋向

21.power-driven *a.* 动力驱动的，电动的

22.external vibrator 表面振捣器

# Lesson 11 Dams

The first dam for which there are reliable records was built on the Nile River sometime before 4000 B. C. It was used to divert the Nile and provide a site for the ancient city of Memphis. The oldest dam still in use is the Almanza Dam in Spain, which was constructed in the sixteenth century. With the passage of time, materials and methods of construction have improved, making possible the erection of such large dams as the Nurek Dam which is being constructed in the Former Soviet Union on the Vaksh River near the border of Afghanistan. This dam will be 1017ft (333m) high, of earth and rock fill. The failure of a dam may cause serious loss of life and property; consequently, the design and maintenance of dams are commonly under government surveillance. In the United States over 30,000 dams are under the control of state authorities. The 1972 Federal Dam Safety Act (PL 92-367) requires periodic inspections of dams by qualified experts. The failure of the Teton Dam in Idaho in June 1976 added to the concern for dam safety in the United States.

## 1 Types of Dams

Dams are classified on the basis of the type and materials of construction, as gravity, arch, buttress, and earth. The first three types are usually constructed of concrete. A gravity dam depends on its own weight for stability and is usually straight in plan although sometimes slightly curved. Arch dams transmit most of the horizontal thrust of the water behind them to the abutments by arch action and have thinner cross sections than comparable gravity dams. Arch dams can be used only in narrow canyons where the walls are capable of withstanding the thrust produced by the arch action. The simplest of the many types of buttress dams is the slab type, which consists of sloping flat slabs supported at intervals by buttresses. Earth dams are embankments of rock or earth with provision for controlling seepage by means of an impermeable core or upstream blanket. More than one type of dam may be included in a single structure. Curved dams may combine both gravity and arch action to achieve stability. Long dams often have a concrete river section containing spillway and sluice gates and earth or rock-fill wing dams for the remainder of their length.

The selection of the best type of dam for a given site is a problem in both engineering feasibility and cost. Feasibility is governed by topography, geology and climate. For example, because concrete spalls when subjected to alternate freezing and thawing, arch and buttress dams with thin concrete sections are sometimes avoided in areas subject to extreme cold. The relative cost of the various types of dams depends mainly on the availability of construction materials near the site and the accessibility of transportation facilities. Dams are sometimes built in stages with the second or later stages constructed a decade or longer after the first stage.

The height of a dam is defined as the difference in elevation between the roadway, or spillway crest, and the lowest part of the excavated foundation. However, figures quoted for heights of dams are often determined in other ways. Frequently the height is taken as the net height above the old river bed.

## 2 Forces on dams

A dam must be relatively impervious to water and capable of resisting the forces acting on it. The most important of these forces are gravity (weight of dam), hydrostatic pressure, uplift, ice pressure, and earthquake forces. These forces are transmitted to the foundation and abutments of the dam, which react against the dam with an equal and opposite force, the foundation reaction. The effect of hydrostatic pressure caused by sediment deposits in the reservoir and of dynamic forces caused by water flowing over the dam may require consideration in special cases.

The weight of a dam is the product of its volume and the specific weight of the material. The line of action of this force passes through the center of mass of the cross section. Hydrostatic forces may act on both the upstream and downstream faces of the dam. The horizontal component $H_h$ of the hydrostatic force is the force on a vertical projection of the face of the dam, and for unit width of dam it is

$$H_h=rh^2/2 \tag{11.1}$$

where: $r$ is the specific weight of water; $h$ is the depth of water. The line of action of this force is $h/3$ above the base of the dam. The vertical component of the hydrostatic force is equal to the weight of water vertically above the face of the dam and passes through the center of gravity of this volume of water.

Water under pressure inevitably finds its way between the dam and its foundation and creates uplift pressures. The magnitude of the uplift force depends on the character of the foundation and the construction methods. It is often assumed that the uplift pressure varies linearly from full hydrostatic pressure at the upstream face (heel) to full tail-water pressure at the downstream face (toe). For this assumption the uplift force $U$ is

$$U = r ( h_1+ h_2 ) t / 2 \tag{11.2}$$

where: $t$ is the base thickness of the dam; $h_1$ and $h_2$ are the water depths at the heel and toe of the dam, respectively. The uplift force will act through the center of area of the pressure trapezoid (Fig.11.1).

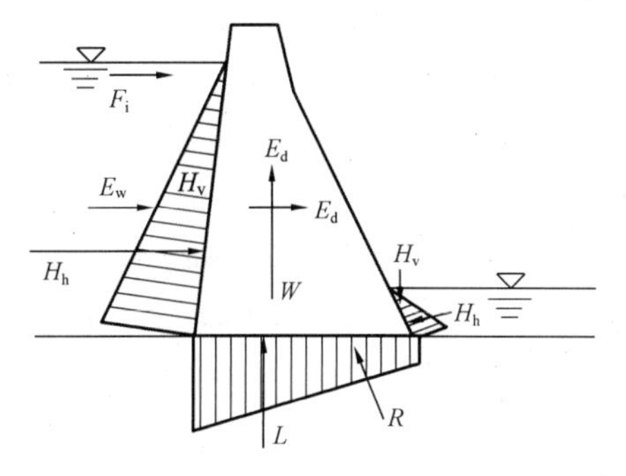


Fig. 11.1 Free-body diagram of the cross section of a gravity dam

Actual measurements on dams (Fig.11.2) indicate that the uplift force is much less than that given by Eq. 11.2. Various assumptions have been made regarding the distribution of uplift pressures.

The U. S. Bureau of Reclamation sometimes assumes that the uplift pressure on gravity dams varies linearly from two-thirds of full uplift at the heel to zero at the toe. Drains are usually provided near the heel of the dam to permit the escape of seepage water and relieve uplift.

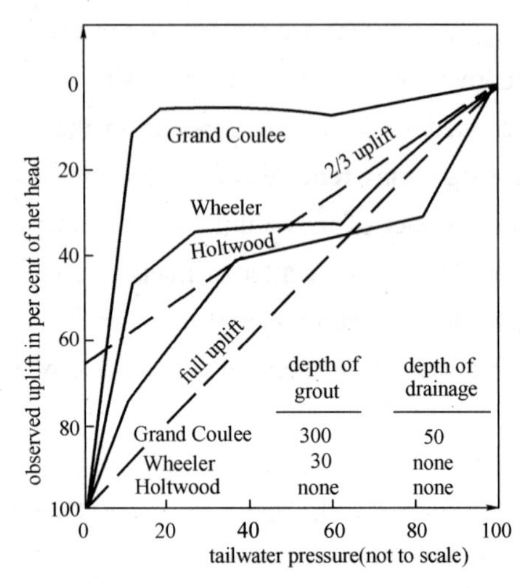


Fig. 11.2 Maximum observed uplift pressures under some existing gravity dams
(Data from the ASCE Committee on Uplift in Masonry dams)

## New Words and Expressions

1.reliable *a.* 可靠的
2.erection *n.* 建筑；安装
3.border *n.* 边沿；边界，田埂
4.consequently *ad.* 从而；因此
5.surveillance *n.* 监视
6.periodic *a.* 周期的，定时的
7.inspection *n.* 检查；调查
8.qualified *a.* 有资格的；合格的，胜任的
9.thrust *n.* 推力
10.abutment *n.* 座；坝座
11.comparable *a.* 比较的；类似的
12.canyon *n.* 峡谷
13.slab *n.* （平）板；（厚）片
14.flat *a.* 平坦的
15.embankment *n.* 土堤；填土堤；堤防
16.seepage *n.* 渗透（漏）；渗出
17.impermeable *a.* 不渗透的；不透水的；密封的
18.core *n.* 心墙；坝心
19.blanket *n.* 铺盖，覆盖层
20.remainder *n.* 剩余，余部；余项
21.topography *n.* 地形，地势
22.geology *n.* 地质，地质学
23.spall *vt, n.* 碎裂；剥落
24.thaw *vt.* 融化；解冻
25.accessibility *n.* 接近性；可达性
26.excavate *vt.* 开凿；挖掘
27.quote *vt.* 引用；引证
28.impervious *a.* 透不过的
29.hydrostatic *a.* 静水（力）的；流体静力（学）的
30.uplift *n.* 扬压力，上托力；上举
31.sediment *n.* 沉积（物）；泥沙
32.inevitably *ad.* 不可避免地
33.heel *n.* 坝踵；上游坡脚
34.tailwater *n.* 尾水；尾水位
35.toe *n.* 坝趾；坡脚，坝脚
36.trapezoid *a.* 不规则四边形（的），梯形（的）

37.assumption *n.* 假设
38.relieve *vt.* 减轻，解脱（开）
39.B. C. =before Christ 公元前
40.the passage of time 时间的推移
41.cross section （横）截面；剖面
42.curved dam 弧形坝
43.sluice gate 泄水闸门
44.rock-fill wing dams 堆石翼坝
45.spillway crest 溢洪道项部；溢流堰项
46.sediment deposits 泥沙沉积；成层沉积
47.horizontal component 水平分力
48.vertical projection 垂直投影
49.seepage water 渗水
50.the Nile River 尼罗河
51.Memphis 孟菲斯城（古埃及城）
52.the Almanza Dam in Spain 西班牙的阿尔曼扎坝
53.the Vaksh River 瓦赫什河
54.Afghanistan 阿富汗
55.the 1972 Federal Dam Safety Act 1972rh （美国）联邦大坝安全法
56.the Teton Dam in Idaho 爱达荷州的泰托坝
57.the U. S. Bureau of Reclamation 美国垦务局

# Reading Material Dam

Large modern dams are constructed across rivers to store and control vast quantities of water. And it can be used for irrigation, electric power generation, flood control, navigation, public water supplies, industrial water supplies, and recreation. Some dams serve only one or a few of these purposes, but most modern dams serve several and therefore are called multipurpose dams.

Based on data from more than 61 countries, the World Register of Dams for dams higher than 50 feet shows there are more than 10,000 dams in the world, and more than 250 dams are being added annually worldwide. The United States builds about 100 dams a year that are taller than 50 feet; Japan builds about 30 such dams annually; and India, Italy and Spain each build about 10 such dams annually. During 1963-1965, the Former Soviet Union built 12 such dams, for an average of 4 dams annually.

In recent times, the demands on water resources caused by expanding populations and economies have increased the need to build dams for water storage. In the United States, for water diverted into waterworks structures, the average water consumption per capita was 137 gallons per day in the late 1960's. This figure includes residential, commercial, and industrial use of water in a community. In towns where there are large industrial users of water, the water requirement per capita is even higher because, for example, it takes 65,000 gallons of water to make a ton of steel and 10 gallons of water to refine a gallon of gasoline.

Although dam building is a key means for controlling and using water resources for the benefit of mankind, it sometimes presents problems because a dam blocks the river, and impedes navigation by vessels and migration by fish. Where a channel for river vessels is required, navigation locks are built alongside the dam. Where fish migration is important, fish locks similar to locks for vessels are provided. Fish ladders also are built so that valuable fish, such as salmon, can swim around a dam. The fish enter at the lowest of a series of ascending pools and leap from pool to pool until they get to the reservoir level.

An advanced technology is needed in building a large dam and its related facilities so that they will be economical, useful, and safe from failure. Aerial photographs often are used in the preliminary stages of exploration to locate major geological features at a site. Expert knowledge of hydrology, geology, hydraulics, and other sciences is applied to solve the complex problems encountered in analyzing the foundation rock and soil, laying bare the riverbed at the dam site, and designing and constructing the dam and related facilities. Each material, such as earth, rock or concrete, to be used in building the dam is selected to produce the desired results in the most economical way at a particular site. Careful analysis and study of each material are made to ensure that the dam will be impervious to water and will be heavy and strong enough to resist the reservoir water pressure against the face of the dam. In high dams, this pressure is tremendous; at the base of the Hoover Dam, for example, the water exerts a pressure of 45,000 psi.

## New Words and Expressions

1.construct *vt.* 建造
2.multipurpose *a.* 多功能的，综合利用的
3.register *n.* 记录表
4.divert *v.* 水分，引水
5.capita （caput 的复数）头
6.residential *a.* 住宅的，居住的
7.requirement *n.* 需要量
8.block *v.* 封锁，截断
9.impede *v.* 阻碍，妨碍
10.migration *n.* 迁移，流动
11.alongside *prep.* 靠在…旁边
12.salmon *n.* 鲑鱼
13.ascending *a.* 向上的
14.leap *v.* 跳过，跃进
15.aerial *a.* 航空的
16.hydraulics *n.* 水力学
17.riverbed *n.* 河床，河底
18.tremendous *a.* 极大的
19.psi （作 p. s. i）（pound per square inch 的缩写）磅/英寸$^2$
20.safe from 没有受到…的危险，无遭受…之患
21.aerial photograph 航摄照片
22.be applied to 适（应）用于
23.lay bare 揭开，清掏
24.impervious to water 防水的，不透水的
25.at the base of 在…基部

# Lesson 12 Embankment Dams

## 1 Selection of embankment type

In general, there are two types of embankment dams: earth and rockfill. The selection is dependent upon the usable materials from the required excavation and available borrow. It should be noted that rockfills can shade into soil fills depending upon the physical character of the rock and that no hard and fast system of classification can be made. Rocks which are soft and will easily break down under the action of excavation and placement can be classified with earthfills. Rocks which are hard and will not break down significantly are treated as rockfills.

The selection and the design of an earth embankment are based upon the judgment and experience of the designer and is to a large extent of an empirical nature. The various methods of stability and seepage analyses are used mainly to confirm the engineer's judgment.

## 2 Freeboard

All earth dams must have sufficient extra height known as freeboard to prevent overtopping by the pool. The freeboard must be of such height that wave action, wind setup, and earthquake effects will not result in overtopping of the dam. In addition to freeboard, an allowance must be made for settlement of the embankment and the foundation which will occur upon completion of the embankment.

## 3 Top width

The width of the earth dam top is generally controlled by the required width of fill for ease of construction using conventional equipment. In general, the top width should not be less than 30 ft. If a danger exists of an overtopping wave caused either by massive landslides in the pool or by seismic block tipping, then extra top width of erosion resistive fill will be required.

## 4 Alignment

The alignment of an earthfill dam should be such as to minimize construction costs but such alignment should not be such as to encourage sliding or cracking of the embankment. Normally the shortest straight line across the valley will be satisfactory, but local topographic and foundation conditions may dictate otherwise. Dams located in narrow valleys often are given an alignment which is arched upstream so that deflections of the embankment under pool load will put the embankment in compression thus minimizing transverse cracking.

## 5 Abutments

Three problems are generally associated with the abutments of earth dams: ①seepage, ②instability, and ③transverse cracking of the embankment. If the abutment consists of deposits of pervious soils it may be necessary to construct an upstream impervious blanket and downstream drainage measures to minimize and control abutment seepage,

Where steep abutments exist, especially with sudden changes of slopes or with steep bluff, there exists a danger of transverse cracking of the embankment fills. This can be treated by excavation of the abutment to reduce the slope, especially in the impervious and transition zones. The transition zones, especially the upstream, should be constructed of fills which have little or no cohesion and a well-distributed gradation of soils which will promote self-healing should transverse cracking occur.

## 6 Stage construction

It is often possible, and in some cases necessary, to construct the dam embankment in stages. Factors dictating such a procedure are: ①a wide valley permitting the construction of the diversion or outlet works and part of the embankment at the same time; ②a weak foundation requiring that the embankment not be built too rapidly to prevent overstressing the foundation soils; ③a wet borrow area which requires a slow construction to permit an increase in shear strength through consolidation of the fill. In some cases it may be necessary to provide additional drainage of the foundation or fill by means of sand drain wells or by means of horizontal pervious drainage blankets.

## 7 Embankment soils

Most soils are suitable for use for embankment construction, however, there are physical and chemical limitations, soils which contain excessive salts or other soluble materials should not be used. Substantial organic content should not exist in soils. Lignite sufficiently scattered through the fill to prevent the danger of spontaneous combustion, is not objectionable. Fat clays with high liquid limits may prove difficult to work and should be avoided.

## 8 Compaction requirements

The strength of the impervious and semi-impervious soils depends upon the compacted densities. These depend in turn upon the water content and weight of the compacting equipment. The design of the embankment is thus influenced by the water content of the borrow soils and by the practicable alternations to the water content either prior to placement of the fill or after placement but prior to rolling. If the natural water content is too high, then it may be reduced in borrow area by drainage, or by harrowing. If the soil is too dry it should be moistened in the borrow area either by sprinkling or by ponding and then permitted to stabilize the moisture content before use. The range of placement water content is generally between 2 percent dry to 2 or 3 percent wet of the standard Proter optimum water content. Pervious soils should be compacted to at least 80 percent of relative density.

If necessary, test fills should be constructed with variations in placement water content, lift thickness, number of roller passes and type of rollers. For cases of steep abutment, the fill must be placed in thin lifts and compacted by mechanical hand tampers. All overhangs should either be removed or filled with lean concrete prior to fill placement.

## 9 Types of instruments

The type of instrumentation depends upon the size and complexity of the project. The devices in common use are: ①piezometers; ②surface movement monuments; ③settlement gages; ④inclinometers; ⑤internal movement and strain indicators; ⑥pressure cells; ⑦movement indicators at conduit joints and other concrete structures.

## New Words and Expressions

1.borrow *v., n.* 借（用，入），采料场，取样，割窃
2.earthfills 填土
3.rockfill 堆石
4.freeboard 超高
5.wind setup 风力壅高
6.massive *a.* 大规模的，大量的，笨重的；~landslide 大规模塌方
7.tip *v.*（使）倾卸（斜），倾翻，倒出，崩（卷，伤）刃
8.block *n.* 块体，砌块，街区，砾石
9.alignment *n.* 定线（位，向），对中，定中心
10.cracking *n.* 破裂，裂缝
11.abutment *n.* 坝座，桥台，岸墩
12.deposit *n.* 沉积物，淤积，矿床（层），存款
13.impervious blanket 不透水铺盖
14.bluff *n.* 陡壁，悬崖
15.transition zone 过渡区
16.well-distributed gradation 颗粒级配良好
17.self healing 自行愈合
18.stage construction 分期施工
19.consolidation *n.* 固结，凝固，联合
20.sand drain well 排水砂井
21.drainage blanket 排水铺盖
22.soluble *a.* 可溶解的，可解的，溶性的；~ materials 可溶物质
23.lignite *n.* 褐煤
24.spontaneous *a.* 自然产生的，自然的，天然的，自发的
25.combustion *n.*燃烧，氧化
26.compact *v.* 压实，（加）压（模）塑
27.content *n.* 内容，容量（积度），含量；water ~，含水量
28.placement 填放，填筑，方位，安排
29.rolling 滚动（轧，压）辗压
30.harrow *n., vt.* 耙子，旋转式碎土机；耕地（平，松）
31.sprinkling *n.* 喷洒，洒水，少量，点滴
32.ponding *n.* 积水（库），（挖池）蓄水，坑尘
33.lift *n.* 填筑层，浇筑层
34.lift thickness 填筑层厚度
35.number of roller passes 碾压遍数
36.tamper *n.* 夯（具，锤，板），批夯机
37.overhang *n.* 突出（物，部分），伸出物
38.lean *a.* 贫（瘠，乏，弱）的，含量少的，质劣的
39.lean concrete 贫混凝土
40.piezoeter *n.* 压力计，测压计
41.gage *n.*（测量）（仪）表
42.settlement *n.*沉陷，解决，确定
43.inclinometer *n.* 侧斜仪，倾角计，量坡仪，井斜仪
44.strain *n.* 应变，拉紧，气质
45.indicator *n.* 指示器（表），显示器，千分表
46.monument *n.* 纪念物（碑，像，馆），标石，界碑
47.silty *a.* 粉（土）质的，粉（粒）的，淤泥的
48.core *n.* 心墙，样（岩）芯，线心，心板
49.pumiceous *a.* 浮石的，（像）轻石的，轻石质的
50.crush *v., n.* 轧碎，压（制）服，砂眼
51.basalt *n.* 玄武岩，黑色磁器
52.spall *n.* 碎片，碎石
53.oversize 超径
54.W. S.= water surface 水面，水位
55.camber *n.* 预留部分（高度）
56.berm *n.* 小平台，马道，钱道
57.ledge *n.* 突出部分

## Reading Material Earth Dams

When an earth dam is built on a pervious foundation, a watertight underground wall, called a cut-off, often is built into the foundation to minimize water seepage under the dam. The cut-off is often built down through the overburden (loose material overlying the rock) and founded on rock. Sometimes the trench for the cut-off is dug to a depth more than 100 feet (30 meters).

Earth dams include those that are mainly built of one material, called homogenous fill. The earth dams also include those that are built with zones of different material, called zoned fill. In homogenous-fill dams, the upstream face is made impervious either by a covering of compacted earth and rock fragments or by a covering of concrete. In zoned fill dams, the innermost zone, or core, is a fine impervious fill that prevents seepage through the dam. Nevertheless, the outmost layers are made of sturdy but not watertight rock.

One method for building earth dams is called the rolled-fill method. In this method, layer on layer of material is placed at the dam site, and each layer is packed down solidly by rollers. The highest completed earth-fill dam is the Oroville Dam (1967) in northern California. This dam has been compacted earth and rock and several zones of different impervious materials. This dam, which is 770 feet (235 meters) high and contains 80.3 million cubic yards (61 million cu meters) of earth and rock fill, is a key structure in the largest water project in history—a $2.8 billion project to supply water to the San Joaquin Valley and southern California.

A second method for building earth dams is called the hydraulic-fill method. In this method, the upstream and downstream portions of the dam are formed, and then coarse particles, silt and clay are sluiced into the large basin between these portions. The hydraulically filled material is allowed to settle, and then the water that carried the material to the basin is drained off. The largest dam built by the hydraulic-fill method is the Fort Peck Dam (1940) in Montana. The dam contains 125.6 million cubic yards (96 million cu meters) of earth and is 250 feet (76 meters) high.

Big earth dams, such as Oroville and Fort Peck, lead by far all other types of dams in terms of volume of material in the dam. Modern earth dams are bigger and higher than earlier ones as a result of recent advances in the scientific knowledge of the behavior of soils. Also, with the use of modern mechanized earth-moving equipment, a big earth dam can be built at a cost that is competitive with the cost of other types of dams.

## New Words and Expressions

1.watertight *a.* 防渗的，不透水的

2.minimize *v.* 将…减至最低程度

3.seepage *n.* 渗漏

4.overburden *n.* 覆盖层

5.overlying *a.* 上覆的

6.trench *n.* 沟槽

7.homogeneous *a.* 类的

8.zoned *a.* 分区的

9.compact *v.* 压实

10.innermost *a.* 最里面，最深处

11.outermost *a.* 最外面，最远的

12.roller *n.* 压路机

13.sturdy *a.* 坚固的

14.sluice *v.* 冲洗，用水力法掘土

15.competitive *a.* 竞争的

16.the Oroville Dam 奥罗维尔坝

17.the Fort Peck Dam 佩克堡坝

18.Montana *n.* 蒙大拿（美国西北部一州）

19.rolled-fill method 辗压式方法

20.hydraulic fill method 水力冲填法

21.lead…by… 比…超前

22.in terms of 根据，就…来说

23.competitive with 与…不相上下

# Lesson 13 Concrete Gravity Dams on Rock Foundations

The designer of any dam must make basic assumptions regarding site conditions and their effects on the proposed structure. Site investigations provide the engineer with much of the information to evaluate these assumptions, the bases for safe dam design. Some important assumptions for small dam design involve uplift pressure, seepage control measures, channel degradation and downstream toe erosion, foundation conditions, and quality of construction. Additional assumptions should involve silt loads, ice pressures, earthquake accelerations, and wave forces.

## 1 Safety Factors

Safety factors should be considered in the light of economic conditions. Large safety factors result in a more costly structure; however, low safety factors may result in failure, which could also lead to high cost. Proper safety factors result only from an adequate determination of sliding, overturning, and overstressing forces within and acting on the dam.

## 2 Overturning

Ordinarily, the safety factor against overturning is between 2 and 3. In smaller dams it is often larger. If the computed safety factor falls below 2, the section of the dam should be modified to increase the safety margin. A gravity dam rarely fails from overturning since any tendency to overturn provides greater opportunity for a sliding force to create the failure. The safety factor against overturning is the ration of the righting moment to the overturning moment about the toe of the dam. This can be expressed as

$$FS_0 = \frac{W_c \times l_1 + W_w \times l_2}{P \times l_3 + U \times l_4}$$

which: $W_c$ is force due to weight of concrete; $W_w$ is force due to weight of water on inclined surfaces; $P$ is force of water acting to displace dam downstream; $U$ is uplift force; $l$ is length of moment arm for respective forces.

Also, if the uplift pressure at the upstream face exceeds the vertical stress at any horizontal section without uplift, the uplift forces greatly increase the tendency for overturning about the downstream toe at that assumed horizontal plane. The dam may still be considered safe if the tension stresses developed are less than the allowable stresses in the concrete and in the foundation material. This assumption is based on good workmanship and development of a tensile strength within the structure on all horizontal planes.

## 3 Sliding

Three approaches are used by engineers in evaluating the safety of a dam from being displaced downstream. Each has merit and generally involves the same relationship of forces. Although the computed values are safe, they are considerably different. The three approaches are: ①a safe sliding factor; ②a safety factor; ③a shear-friction safety factor. Clear distinction must be made among these three approaches. The primary purpose of each is to obtain a safe coefficient that when exceeded would put the dam in jeopardy of being pushed downstream.

The sliding factor of a gravity dam with a horizontal base equals the tangent of the angle between the perpendicular to the base and the resultant foundation reaction. The sliding factor for small dams is computed by taking the ratio of the summation of horizontal forces $\Sigma P$, to the summation of vertical forces $\Sigma W$, including the uplift $U$, or

$$f = \tan\theta = \frac{\Sigma P}{\Sigma W - U}$$

If $f$ computed in this manner is equal to or less than the static friction coefficient $f'$, the dam is considered safe. A unit width of 1 ft. is assumed for these calculations. Safe values for the sliding factor coefficient are given in Table 13.1 for various foundation materials.

Table 13.1 Allowable sliding factors for various foundation conditions

| Material | Safe sliding factor, $f$ | Suggested minimum factor of safety, $f_s$ | Shear sliding factor, $SSF$ |
|---|---|---|---|
| Concrete on concrete | 0.65~0.8 | 1~1.5 | 4 |
| Concrete on sound rock, clean and irregular surface | 0.8 | 1~1.5 | 4 |
| Concrete on rock, some laminations | 0.7 | 1~1.5 | 4 |
| Concrete on gravel and coarse sands | 0.4 | 2.5 | — |
| Concrete on sand | 0.3 | 2.5 | — |
| Concrete on shale | 0.3 | 2.5 | — |
| Concrete on silt and clay | * | 2.5* | — |

* Tests required to determine safety.

The factor of safety, $f_s$, against sliding is defined as the ratio of the coefficient of static friction $f'$, to the tangent of the angle between a perpendicular to the base and the direct foundation reaction, expressed as

$$f_s = \frac{f'}{\tan\theta} = \frac{f'(\Sigma W - U)}{\Sigma P}$$

This approach also assumes shear forces as added safty measure. The safty factor against sliding is usually between 1 and 1.5 for gravity dams on rock utilizing a conservative cross-section. The inclusion of uplift and seismic forces in the calculations may reduce the safty factor to about unity. These values are for safety against sliding on a horizontal plane; if the foundation slopes downstream, the safety factors against sliding on the plane of the base are correspondingly reduced. Designers often use concrete placed into cutoffs or rock foundations to decrease the sliding tendency of the dam.

Another approach, favored by many engineers, includes the evaluation of shear into the safety factor. The shear-friction relationship is:

$$SSF = \frac{f'(\Sigma W - U) + b\sigma}{\Sigma P}$$

which: $b$ is base length at plane of shear being studied; $\sigma$ is allowable working shear stress of material or materials at plane of shear.

Safety factors computed in this manner should approach values used in normal structural computations. Static friction values often assumed for concrete moving rock or concrete on concrete surfaces varies from 0.65 to 0.75. The working shear stress $\sigma$ of concrete is related to the compressive strength of concrete. The unit shearing strength of concrete is about one-fifth of the compression breaking stress from standard cylinders. This indicates a strength of 400 to 800 psi for concrete in dams. It also provides a safety factor of 4 if the unit working stresses used in computations are 100 to 200 psi. Greater working stresses are not recommended unless the concrete for the smaller dams is actually pretested.

## 4 Internal Concrete Stresses

The unit stresses in the concrete and foundation materials must be kept within prescribed maximum values to avoid failures. Small dams normally develop stresses within the concrete that are less than the actual strength that may develop if the proper concrete mix is used. A concrete mix that ensures durability will normally have sufficient strength to provide an adequate safety factor against overstressing.

Overstressing the foundation material must also be investigated. In small dams this is pertinent in jointed rocks and soft foundations such as gravel and sand. The designer should check the local codes for allowable bearing pressures and confer with qualified engineers to evaluate the foundation materials. Table 13.2 provides suggested allowable bearing values for initial studies and guides in designing small concrete dams.

Table 13.2 Weighted creep ratios load-bearing values of foundation materials

| Material | Lane's weighted creep ratio | Bligh's coefficients | Allowable bearing values（t/ft$^2$） |
|---|---|---|---|
| Very fine sand or silt | 8.5 | 18 | 3 dense |
| Fine sand | 7.0 | 15 | 1 loose |
| Medium sand | 6.0 | — | 3 |
| Coarse sand | 5.0 | 12 | 3 |
| Fine gravel | 4.0 | — | 5 |
| Medium gravel | 3.5 | — | 5 |
| Gravel and sand | 3.0 | 9 | 5~10 |
| Coarse gravel, including cobbles | 3.0 | — | 5~10 |
| Boulders with some cobbles and gravel | 2.5 | — | 10 |
| Boulders, gravel, and sand | — | 4.6 | 5 |
| Soft clay | 3.0 | — | 1 |
| Medium clay | 2.0 | — | 4 |
| Hard clay | 1.8 | — | 6 |
| Very hard clay or hardpan | 1.6 | — | 10 |
| Good rock | — | — | 100 |
| Laminated rock | — | — | 35 |

## New Words and Expressions

1.degradation *n.* 降低，减少，河底，刷深 channel ~ 河槽刷深
2.margin *n.* 余量 [额，地]，裕[幅] 度，页边
3.moment *n.* 力矩，动量；overturning ~ 倾覆力矩；~ arm 力臂
4.measure *n.* 量度，测量，度量，措施，方法，程度
5.cross-section 横断面
6.shear *v., n.* 剪切，剪（应变）
7.stress *n.* 应力，胁迫，强调；tension ~ 拉应力；breaking ~ 破坏应力
8.pretest *n.* 事先试验，预先检验；*a.* 试验前的；field test 野外试验，田间试验
9.joint *n., vt.* 节理，结点，接缝，连接
10.sound *a.* 健全的，坚固的，彻底的，合理的
11.allowable stress 允许应力
12.workmanship 工作质量，（制造）工艺，技巧
13.tensile *a.* 拉[张]力的，抗拉的，可引伸的
14.factor *n.* 系数，倍（数），商，因素
15.friction *n.* 摩擦，摩阻，冲突
16.static *a.* 静（止，位，态，力，电，压）的，固定的
17.coefficient *n.* 系[因，常]数，折算率，程度
18.mix *n.* 混合，混合料，配合比
19.weighted *a.* 加[计]权的，权重的，受力的，已称重的
20.creep *n.* 渗（水），漏电，蠕变，滞缓，爬行
21.ratio *n.* 比（率，例，值），传动比，（变换）系数，关系
22.load-bearing value 承载值
23.lamination *n.* 层压（成型），层叠（合，理），纹理
24.shale *n.* （油）页岩，板岩
25.boulder *n.* 漂[巨]砾，圆[五，卵，拳，砾]石
26.cobble *n.* 圆石，（大，鹅）卵石，中砾石，铺路石
27.hardpan *n.* 硬土层，硬质地层，硬盘，底价
28.cylinder *n.* 圆柱（体）汽（气）缸，圆筒，柱面
29.in the light of … 根据，按照，鉴于…
30.confer with … 与…商量
31.be defined as … （被）定义为…
32.be expressed as … （被）表达为…

## Reading Material Concrete Gravity Dams on Soft Foundations

Dams on soft foundation materials must be safe against the same forces as dams on good rock foundations. In addition, the designer must consider the effects of seepage, piping underneath the dam, and subsidence or consolidation of the foundation materials.

From an engineering viewpoint, the water movement below or through a dam is not objectionable if the flow does not exceed safe limits of design. However, the total flow may be large enough to make it economically desirable to seal off the permeable zone. Water that passes under the dam is lost for that project; however, stream flow may be sufficient for the designer to reduce construction costs by permitting some flow within safe limits. These limits may be summarized for pervious dam foundations as:

(1) Limit percolation velocities and pressures in the foundation material so they do not move the soil particles to cause piping, scouring of the foundation or mass flow of the material.

(2) Limit the seepage uplift pressure under the foundation so that no undesirable overturning

moment and no sliding on the foundation occur.

## 1 Flow in Porous Material

Flow through porous material may be estimated by using the Darcy equation:

$$Q = kiA$$

where: $Q$ is discharge in a given unit of time; $k$ is coefficient of permeability for the foundation, i. e., discharge through a unit area at unit hydraulic gradient; $i$ is hydraulic gradient, difference in head divided by length of path, h/L; $A$ is gross area of foundation through which flow takes place.

The coefficient, $k$, is determined by several methods. The most satisfactory and economical for small dams is the pumping in test. In this test, water is pumped into a drill hole or test pit; the seepage rate is observed under a given water head.

The pumping-out test is relatively expensive and results are more difficult to interpret. This test measures gross permeability by pumping water from a well hole at a constant rate and measuring the draw-down of the water table in observation wells at various distances from the pumped well.

The dye test relates the rate of flow of a dye or electrolyte from the point of injection to an observation well. This test requires several trials since the assumed direction of flow for the dye may or may not be along the underground flow paths. Several trials (with observation well relocations) may be needed to adequately estimate permeability.

## 2 Seepage Forces

The magnitude of the seepage forces through the foundation and at the downstream toe of the structure depends on the rate of head loss of the moving water. Impervious soils are not as susceptible to piping because they offer greater resistance to flow, and the reservoir head is largely dissipated by friction. On the other hand, pervious soils (and stratified or fractured rock) may permit substantial flow movement at the downstream toe of the dam without much loss of head due to friction. In such instances, the designs must be investigated to ensure against blowouts.

Another type of failure is due to internal erosion from springs near the downstream toe. It proceeds upstream along the base of the dam, the walls of a conduit, a bedding plane, or other type of weakened zone. This type of failure is due to subsurface erosion.

The magnitude and distribution of seepage forces may be determined by a flow-net analysis. A flow net is a graphic representation of percolation paths and lines of equal potential (pressure plus elevation above a datum) in subsurface flow. The flow net is limited in the analysis of problems of stratified flow or problems of subsurface spring erosion failures. Further, drawing an accurate flow net requires considerable experience, particularly when cutoff walls are used and non-homogeneous soils make up the foundation.

Incipient piping occurs when the pressure exerted on the soil by moving water exceeds the resistive force offered by the soil. The seepage force, $P_f$, exerted on the soil by the water, equals the unit weight of water, $w$, times the hydraulic gradient, $i$, which acts on a unit volume of soil. If the soil is homogeneous, the force acts uniformly on the whole soil mass. The critical point for incipient piping is at the downstream toe of the dam.

Darcy's law permits a theoretical basis for providing an adequate length of water travel beneath a dam. The relationship for discharge may be stated as

$$Q = kiA = C_t HA/L$$

where: $Q$ is discharge, cfs; $H$ is head (reservoir depth), ft; $L$ is length of path, ft; $C_t$ is coefficient dependent on the material.

From continuity, $Q=AV$; hence, $L=C_t H/V$.

For a given material, there is a maximum velocity, V, at which water can emerge from below the dam without causing failure by carrying away the foundation material. The length of travel is established by the weighted line of creep, as proposed by Lane. Lane's procedure places greater emphasis on vertical travel than on the horizontal paths (in contrast with Bligh's equal weight values). In computing the line of creep for Lane's procedure, the horizontal distances of flow are taken as one-third the value of vertical distances. Vertical distances, including 45-deg slopes, are taken as full-length values for creep distance.

One point not fully evaluated in either procedure concerns incipient piping below the dam. Both authors recognize the increased hydraulic gradient at the downstream toe by requiring that cutoffs and drainage filters be properly placed at the toe. A safe pressure gradient must be ensured at this critical point in design. The more rapid the upward reduction in pressure, the less stable the material, until at a critical value the material actually mover or floats out. This critical value is expressed as

$$H/L=(S-1)(1-P)$$

where: $S$ is specific gravity of material; $P$ is percentage of voids in material expressed as a decimal.

Safety from toe flotation can best be accomplished by providing a cutoff wall and inverted sand-drain filter upstream of the wall, with pipe drains carrying the water to the surface. These effectively reduce uplift pressures and eliminate piping. This type of cutoff should be made in conjunction with an upstream cutoff wall that reduces uplift forces.

The electric analogy is an economical, rapid method to analyze cutoff walls and dam foundation conditions. This method makes use of Darcy's law to develop a flow net that is analyzed for the various forces and critical conditions of flow. Flow nets can be constructed by a mathematical method, graphic method, or model experiments.

## New Words and Expressions

1.subsidence *n.* 下沉（陷），沉（陷，降没）

2.consolidation *n.* 巩固，固结，联合

3.resistance *n.* 阻力，（抵）抗力，抗[耐]…性 ~to sliding 抗滑能力

4.percolation *n.* 渗透[滤，漏]（作用），深层渗透，地面渗入

5.alluvial *a.* 冲[淤]积的，冲积土[层；矿床]，淤积土

6.scour *v.* 擦洗，洗刷，冲洗，疏浚

7.Darcy equation 达西公式

8.mass flow 质量流（量）

9.hydraulic gradient 水力梯度

10.gross *n.* 总（毛）重，全体，总数

11.media *n.* medium 的复数，介质，中间（物），适度

12.blowout *n.* 管路清除，风力移动，贫矿脉，熔解，吹（喷）出，井喷

13.bedding *n.* 铺盖，层理[面]，基床[底，坑]; bedding

plane 岩层面

14.weakened zone 软弱带

15.seepage force 渗流力

16.flow line 流线

17.incipient *a.* 升（起，原）始的，最初的，初期的；~piping 初期管涌

18.cfs 立方英尺/每秒

19.specific *a.* 比（率），比较的，单位的

20.void *n.* 孔隙，孔率，空洞

21.drill *n.* 钻孔，凿片，训练

22.pumping-out test 抽水试验

23.drawdown *n.*（水位）下降，降落，消耗，减少，收缩

24.dye *v.* 染（色），着色；*n.* 染料[液，剂，色]

25.electrolyte *n.*电解（溶）液，电解（离）质

26.injection *n.* 注射[入]，喷射，充满

27.stratified *a.* 有层次的，分层的，成层的；~ flow 层流

28.homogeneous *a.* 均匀[质，相，一]的，同质[性，族，类，种]的

29.nonhomogeneous 非均质的

30.flow net 流网

31.percolation path 渗径

32.line of equal potential 等势线

33.equipotential line 等势线

34.datum *n.* 数据，基准（点，线，面），已知（条件）

35.flotation *n.* 漂浮（性），浮（力），浮动（性），设立，创立，发行（债券）

36.invert *v.* （使）颠倒，（使）倒转，使反向；inverted sand-drain filter 砂质排水反滤层

37.analogy *n.* 模拟（比），类推（比），类似

38.graphic *a.* 图（解，式，形）的，绘图的

# Lesson 14 Arch Dams

Ideally suited to narrow canyons composed of rock, the arch dam provides an economical and efficient structure to control the stream flow. The load-carrying capacity of an arch dam enables the designer to conserve material and still maintain an extremely safe structure.

## 1 Theory

The general theory of arch dam design is comparatively new and changing rapidly as more information is obtained. Engineers have cautiously applied mathematical theory, the law of mechanics, and theories of elasticity to reduce the thickness of arch dams and gain substantial economies.

Historically, the method of analysis of arch dams grew from the cylinder theory to independent arches, crown cantilever analysis, and then multiple cantilever analysis.

The crown cantilever and arch action theory has proved satisfactory for arch dam analysis. In this analysis the horizontal water load is divided between the arch rings and vertical cantilevers. The calculated deflections in the arch and cantilever at conjugate points in the dam are adjusted by changing the dam's shape. This provides nearly equal elastic deflections. Stresses calculated when the deflections are in satisfactory agreement are considered as true stresses in the dam. The shape or form of the dam changes the load distribution. Modification of the shape is by trial and error. Consequently, the procedure is called the trial-load method of analysis.

Another procedure being studied involves the application of the shell theory to arch dam analysis. Most of the simultaneous equations for analysis of a dam can be quickly established except those dealing with irregular boundary conditions. The solution of the simultaneous equations is accomplished by an electronic computer.

The influence of temperature change may be small in flexible arches but is important in flat, thick arches. Further, both temperature and shrinkage effects may vary throughout the thickness of the arch.

Efforts to minimize temperature stresses are logically aimed at construction and design techniques. The designer selects the arch form with consideration of temperature change to minimize internal tension stresses.

Overturning and sliding are seldom a factor in the design of arch dams. The designer's principal interest is in shear, compressive, and tensile stresses. Detailed analyses and computer programs are available. The Corps of Engineers, Bureau of Reclamation, and several consulting firms have computer programs.

## 2 Basic Guides

Previous design experiences with arch dams will assist the designer. In the preliminary selection of an arch dam, skill is needed to arrange the several structures and related facilities to gain economy and not jeopardize the basic safety of the dam.

The arch dam is economical in narrow canyons where the top chord length is about two to three times the height of the dam. Wider valleys require increasing thickness of arches to maintain suitable stresses. Arch dams have competed economically and structurally with other dams for length-to-height ratios well above 5.

The top arch ring thickness usually depends on durability, necessity for a roadway, or ice pressures. A general guide is to maintain the arch thickness at any level, equal to or greater than about 0.02 $R$, where $R$ is the centerline radius for the arch. A minimum thickness value is necessary to guard against buckling.

Economy is gained by making the central angles as large as possible; theoretically, a central angle of about 133deg. is most advantageous. The angle at which the arch ring approaches the rock abutment should be not less than 30deg. Angles greater than 45deg, are permissible, but load-carrying arch action is increasingly sacrificed to cantilever action. This in turn requires thicker arch shapes to maintain stresses within allowable limits.

For safety, designers have increased the allowable design stresses to 1,000psi in compression and 150psi in tension when concrete strengths of 3,000 to 4,000psi in 90 days are required. Through a better understanding of structural behavior, it is realized that the safety of an arch dam is not merely equal to the cylinder strength divided by the allowable design stress. Reliable tests on models have confirmed analytical findings that the safety factor will seldom be less than twice the ratio of cylinder strength to allowable stress. Thus, a safety factor of 9 to 14 actually results and will probably be reduced as designers gain more knowledge of arch action and abutment reactions.

## 3 Foundation Problems and Treatment

The abutment is the foundation of an arch dam and must carry the loads delivered by the arches —the shear, thrust and bending moments. Shear is seldom excessive. Thrust varies with elevation and is usually maximum at mid-height. Thrust is evaluated by summing the stresses across a section. Bending moments at the abutments result in uneven stress distribution that may produce localized stresses in excess of the allowable stress. Allowable stresses at the abutment are the maximum permissible stresses for either the rock or the concrete, whichever is less. Excessive stresses at the abutment may be reduced by reshaping the arch at the elevation where the excess occurs or by thickening the arch at the abutment. Enlargement of the bearing area may be used to reduce stress intensities and can be accomplished by the addition of fillets on the intrados. Reshaping is preferred where possible, since this has least effect on concrete quantities.

Abrupt widening of the valley near the top of the dam may require thrust blocks to transfer arch thrust to the foundation.

Elastic properties of the foundation rock will cause some yielding, which redistributes stresses in the structure. The elastic properties of rock are considered in the crown cantilever analysis method.

## 4 Structural Models

Structural model tests provide a realistic design approach for arch dams. Properly conducted, model studies can predict the stresses in the dam as confirmed by stress measurements in the prototype after construction.

Models make it possible to consider the asymmetry of the dam, the use of a free or perimetrical joints and radial joints, and irregular foundation and abutment surfaces. The model data have led to refinements in mathematical analysis. Likewise, electronic computers have been programmed to reduce time-consuming mathematical analysis, and, in fact, to provide mathematical models.

## New Words and Expressions

1.law of mechanics 力学定律
2.crown *n.* 冠（顶），拱顶，轮周
3.cantilever *n.* 悬臂（梁）伸臂，支撑木
4.crown cantilever analysis 拱冠悬臂梁分析
5.multiple cantilever analysis 多悬臂梁分析
6.deflection *n.* 挠度（曲），偏差，倾斜
7.conjugate *a.* 共轭的，偶合的，成对的
8.trial *n.* 试（验，用，探，算，车，飞，运转），近似解，审讯；~and error (method) 试算法
9.shell *n.* 壳（体，层）外壳
10.simultaneous *a.* 同时（存在，发生，做出）的，联立（方程）的
11.chord *n.* 弦，可变基准线，琴线
12.debris *n.* 碎片[石]，垃圾，有机物残渣
13.analytical findings 分析成果
14.boundary condition 边界条件
15.finite difference 有限差公法
16.buckling *n.* （纵）弯曲，屈曲，弯折，扣住，下垂
17.corp= corporation （有限）公司，协会；corps of Engineers 工程师协会
18.consult *v.* 商量（议），顾问，请教，查阅
19.bureau *n.* 局，科，处，司，社，所，大衣柜，写字台
20.reclamation *n.* 开垦，改良，回收，改造 Bureau of Reclamation 垦务局
21.thrust *n.* 推力，冲，猛推
22.reshape *v.* 改造，重新整型[修整]
23.intensity *n.* 强度，密度，应力，强裂
24.fillet *n.* 镶，嵌（木，条，线），（内）圆角
25.intrados *n.* 拱腹[内弧]线，拱内圈，拱里
26.thrust block 推力墩
27.leakage *n.* 漏水，渗漏
28.yielding *a., n.* 塑性变形（的），产生，可压缩性（的）
29.prototype *n.* 原型（体），设计原型，典型
30.joint arrangement 接缝布置
31.asymmetry *n.* 不[非，反]对称（性，现象）
32.symmetry 对称性
33.perimetrical joint 周边接缝
34.radical joint 径向接缝
35.layout *n.* 布置，分布，安排

## Reading Material Arch Dams and Buttress Dams

Arch Dams. A single-arch dam, which has an upstream face that curves from bank to bank in the upstream direction, resists the water pressure of the reservoir chiefly by its arch action. The arch transmits the water pressure both to the abutments at the sides of the dam and to the foundations, thus making use of the compressive strength of the arch shape. As compared with a gravity dam of the same height and width, the principal advantage of the arch dam is that a smaller volume of material is required. This is so because the shape of the dam, as well as its weight contributes to its stability.

Buttress Dames. The main structural elements of a buttress dam are a watertight upstream face and a series of buttresses that support the face and transmit the water pressure and the weight of the structure to the foundation. The upstream face usually is inclined at an angle of about 45 degrees. The buttresses generally are triangular-shaped walls that are spaced either at close or comparatively wide intervals along the backside of the upstream face.

A basic advantage of the buttress dam is that the weight of the water on the inclined face contributes to the stability of the dam and permits a saving in the amount of concrete that otherwise would be required for dead weight. This dam also can accommodate minor foundation movements without serious damage to it.

Except for the difference in shape the number of structural elements is the main difference between the buttress dam and the solid gravity dam. Another difference is that the buttress dam partly relies on the strength of the materials of its structural elements, whereas the solid gravity dam relies on its weight rather than the strength of the material with which it is built. As a consequence, a buttress dam requires only 40% to 80% of the amount of material that a solid gravity dam of similar size would require.

There are several kinds of buttress dams, including the flat slab and buttress dam, the multiple-arch dam, and the hollow gravity dam.

The flat slab and buttress dam, an American dam innovation, has a reinforced concrete slab as its upstream face. The slab, usually sloped at 45 degrees, spans slender buttresses at close spacing. The sloped slab contributes significantly to the stability of the dam because of the weight of the water acting on it. The bottom of the slab is connected to a cut-off (impervious wall) that is built in a trench deep enough to found the wall on rock and thereby prevent leakage under the dam.

The multiple-arch dam has a series of arches that form the upstream face of the dam. Solid or hollow buttresses at the ends of each arch support the face, which generally is built with a slope of about 45 degrees. The arches usually are made of reinforced concrete. The buttresses, generally made of concrete, sometimes are placed at comparatively wide intervals.

The hollow gravity dam, which superficially resembles a hollowed-out gravity dam, is basically a buttress dam because it uses buttresses that act as structural elements and contribute to the stability of the dam. The buttresses are spaced at intervals along the backside of the upstream face to support it. The area between any two adjacent buttresses is a hollow.

## New Words and Expressions

1.abutment *n.* 拱座

2.compressive *a.* 加压的，压缩的；
compressive strength 抗压强度

3.space *v.* （每隔一定间距进行）放置，留间隔

4.accommodate *v.* 使适应

5.minor *a.* 较小的，较轻的

6.multiple *a.* 多重的，复合的
multiple-arch dam 连拱坝

7.hollow *a.* 空心的；
hollow gravity dam 空心重力坝

8.innovation *n.* 革新，改革

9.span *v.* 跨越，横跨

10.slender *a.* 细长的，狭的

11.interval *n.* 间隙

12.superficially *ad.* 表面上，外部地

13.resemble *vt.* 类似，相似

14.as compared with 与…相比

15.contribute to (towards) 有助于；为…出力

16.structural element 结构构件

17.dead weight 自重，净重

18.solid gravity dam 实体坝，重力坝

19.as a consequence 因此，从而

20.flat slab and buttress dam 平板扶壁坝

# Lesson 15 Spillways

A spillway is the safety valve for a dam. It must be designed to discharge maximum flow while keeping the reservoir below a predetermined level. A safe spillway is extremely important. Many failures of dams have resulted from improperly designed spillways or spillways of insufficient capacity. Spillway size and frequency of use depend on the runoff characteristics of the drainage basin and the nature of the project. The determination and selection of the reservoir inflow design flood must be based on an adequate study of the hydrologic factors of the basin. The routing of the flow past the dam requires a reasonably conservative design to avoid loss of life and property damage.

Space limitations do not permit an adequate hydrologic treatment of flood flows. However, data are supplied for estimates of maximum flows for the initial project studies. A more detailed hydrologic analysis is necessary for the utilization of the annual and long-term stream flow in a proper project formulation.

## 1 Stream Flows

The study of stream or river flows involves: ①the determination of the amount of water available throughout a period of years; and ②the determination of the maximum volumes of water that must be handled for spillway design and dam safety.

In the first aspect, the flow is studied for periods of drought and periods of excess water in relation to use in the project development. Prior water rights must be investigated and programmed into this study. A mass curve of the stream runoff over a period of years is developed to determine the available water. The mass curve is the accumulative total of the volume of flow past a given point on the stream over a period of time. Unfortunately, most of the small streams do not have sufficient records to develop the hydrologic information. The engineer usually develops synthetic curves from neighboring stream data and rainfall information. However, methods are available in various texts, journals, and reports to estimate stream flow.

The second aspect involves estimating maximum flood flow to determine spillway requirement and dam safety. Studies show that flood flows are associated with frequency of the event occurring within different time periods. This permits the engineer to make a realistic estimate of the risk of floods causing damage by exceeding the estimated design flow.

If failure of the dam would result in loss of life, the spillway must have sufficient capacity to prevent failure when the maximum probable flood is routed through the reservoir. This is particularly important in rock and earthfill dam that may be overtopped during a flood. Concrete dams can generally withstand some overtopping without failure if the structural analysis adheres to the generally accepted safety factors.

The case of dam failure that does not endanger life may be justified if the organization involved fully realizes the risks and ensuing damages. This situation may exist on low, small reservoir-type dams.

A quick estimate of maximum probable flow can be obtained from a figure. Discharge determined from these curves should be modified by application of hydrologic data pertinent to the area. The curves are based on records of unusual flood discharges for unregulated streams. Creager' gives the equation for the envelope curve in the general form as

$$Q=46CA^{(0.894A-0.48)}$$

or

$$q=46CA^{(0.894A-0.48)-1}$$

which: $Q$ is estimated maximum flood peak, cfs; $q$ is corresponding flood expressed in cfs per square mile of drainage area; $A$ is drainage basin area in square miles; $C$ is coefficient depending on characteristics of drainage area.

The engineer should not accept the flood peak established from these experience curves without first bringing the data up to date to show all recent flood events and those pertinent to the area of study. The coefficient of $C=100$ generally provides a conservative estimate of flood potential for initial design layout.

An advanced procedure to estimate the maximum flood is to transpose storms producing great floods in the region over the drainage basin. The resulting flood is analyzed to determine the peak flow and the flow hydrograph. The hydrograph is the relationship of discharge and time for flood-producing characteristics. A similar approach results from studying the maximum probable precipitation when combined with other flood-contributing characteristics of the basin (including melting snow) to produce the flood hydrograph.

Floods less than maximum may be used for structures where loss of human life is not involved. In minor structures with insignificant storage, where it is permissible to anticipate failure within the useful life of the project, a 50-year or 100-year frequency flood may be used for the inflow design flood.

## 2 Spillway Types

Site conditions greatly influence the location, type, and components of the spillway. The type of dam construction is also influenced by the type of spillway and spillway requirements.

There are six general categories of spillways: ①overflow, ②trough or chute, ③side channel, ④shaft or glory hole, ⑤siphon, ⑥gated. The designer may use one or a combination of types to fulfill the project needs.

Some designs will use one type of spillway for normal operation and for flood peaks up to a 50-year or 100-year frequency storm. An emergency spillway provides additional safety if emergencies arise that was not covered by normal design assumptions. Such situations could result from floods above a certain level, malfunctioning spillway gates, or enforced shutdown of outlet works. The emergency spillway prevents overtopping the main portion of the dam and is particularly needed for earth and rock embankments.

The overflow spillway is well suited to concrete dams. It is commonly used where dams have sufficient crest length for the desired discharge capacity and where the foundation material is solid or can be protected against scouring. Some dams use a free overflow or non-supported type; others incorporate a chute or trough to carry the flow to the downstream channel.

Chute spillways are often used for earth dams or where there are poor downstream foundation materials. Side channels and shaft spillways are readily adapted to narrow canyons where space is limited. Limitations on crest length or maintaining a constant headwater level fit the flow characteristics of a siphon spillway. Gated spillways are used when it is desirable to limit the effects of the dam during high flows and prevent excessive flooding.

The spillway may be part of the dam or a separate structure. Its function must be integrated with the dam. The location, size, and other dam features influence the spillway location and arrangement. The final plan is governed by the overall economy and hydraulic sufficiency of the spillway.

**(to be continued)**

## New Words and Expressions

1.valve *n.* 阀（门），活门，闸门（板）
2.drainage basin 流域
3.runoff characteristics 径流特性
4.reservoir inflow design flood 入库设计洪水
5.period of excess water 丰水季节
6.mass curve 累积曲线
7.routing *n.* 发送（指令），程序安排，调洪演算，调泄
8.stream flow 河川流量
9.50-year frequency flood 50 年一遇洪水
10.side channel 边渠，侧槽
11.shaft *n.* 竖井，旋转轴，柱身，旗杆
12.siphon *n.* 虹吸，虹吸管，弯管
13.emergency spillway 非常溢洪道
14.unregulated stream 未整治的河流
15.envelope curve 包络（曲）线
16.synthetic curve 综合曲线，假想曲线
17.unusual flood discharge 非常洪水泄量
18.hydrologic factor 水文因素
19.melting snow 融雪
20.flood-producing characteristics 洪水产生特性
21.flood-contributing characteristics 洪水形成特性
22.trough *n.* 水槽，盆，地沟
23.chute *n.* 斜槽，瀑布，泻物架，斜管
24.flow (flood) hydrograph 流量（洪水）过程线
25.shutdown 关闭，停工
26.outlet works 泄水[排水，河口]工程
27.dischange capacity 泄水能力

## Reading Material Spillways

**(continue)**

### 3 Spillway control Capacity

The discharge capacity of the spillway control section is determined by the appropriate weir or orifice formula. The uncontrolled crest spillway discharge is given by

$$Q=CLH^{3/2}$$

where: $Q$ is total discharge over spillway, cfs; $L$ is net length of crest, ft.; $H$ is design head on spillway crest, ft.; $C$ is coefficient of discharge.

This equation is modified for piers placed on the crest and for approach flows that retain a considerable velocity component to

$$Q = C\,(L - KNH)\,(H + h_v)^{3/2}$$

where: $K$ is pier contraction coefficient; $N$ is number of pier contractions (2 per pier); $h_v$ is velocity head of approach flow in feet.

In both equations the coefficient, $C$ varies with the configuration of the weir crest and the energy head $H_e$ on the crest. For an ogee spillway, $C$ varies with head, approaching 4.0 as the design head. $H$ is approached. The spillway-rating curve can be computed from the data in Figure, the U. S. Army Corps of Engineers, Waterways Experiment Station, Hydraulic Design Chart .

The depth of approach flow affects the discharge coefficient. Shallow approach depths, Ha, reduce the discharge coefficient from the design coefficient of 4.0 to the low weir coefficient of 3.0. In general, the depth to the weir crest should be equal to or greater than the head acting on the weir. When the approach channel depth is less than the head, boundary conditions affect the flow pattern. In turn, the discharge coefficient approaches the low weir discharge coefficient. Designs where this is important include low diversion dams for water supply or irrigation head works. The velocity is more significant when the approach depth is shallow.

Gated spillway capacities are determined by using the orifice equation

$$Q = C_0 A \sqrt{2gH}$$

or for partial gate openings

$$Q = 2/3\sqrt{2g}CL(H_1^{3/2} - H_2^{3/2})$$

where: $H_1$ is total head in feet on top edge of opening; $H_2$ is total head in feet on bottom edge of opening; $L$ is net length of opening; $C$ is coefficient of discharge.

The coefficient $C$ differs with different gate and crest arrangements. It will be influenced by the approach and downstream flow conditions as they affect the contraction of the flow through the opening. A figure shows coefficients representing averages determined for various approach conditions and downstream conditions for orifice flows at various gate openings to the total head. Coefficients from these figures are sufficiently reliable to determine discharge capacity of spillway structures on small dams.

## 4 Energy Dissipators

Properly designed energy dissipators are essential to dam safety. The flood flows passing the spillway gain kinetic energy that must be properly handled or severe scour can endanger the structure.

The hydraulic jump is used as an effective means of energy dissipation. Other dissipators include plunge pools, ski jumps, and frictional resistance on erosion resistant bed material. In designs utilizing conduits, diffusors have been used with outlet control gates and valves.

All types of stilling basin designs use the principle of the hydraulic jump, which is the conversion of high-velocity flow to velocities that cannot cause damage to the stream channel. The hydraulic jump theory, based on pressure momentum, gives.

$$D_2 = -\frac{D_1}{2} + \sqrt{\frac{D_1^2}{4} + \frac{2V_1^2 D_1^2}{g}}$$

where: $D_2$ is flow depth downstream of the jump, ft; $D_1$ is flow depth upstream of the jump, ft; $V_1$ is velocity in the flow upstream of the jump, ft. per second; $g$ is acceleration of gravity, 32.2 ft. per second$^2$ .

The relationship of depths is often written in terms of the Froude numbe

$$F_1 = \frac{V_1}{\sqrt{gD_1}}$$

as $$\frac{D_1}{D_2} = -1/2 + \sqrt{1/4 + 2F_1^2}$$

This equation shows that the ratio of depths is a function of the Froude number. The stability of the jump is greatly influenced by the Froude number. A design summary of stilling basin characteristics can be provided in figures (omitted).

A chart of stone size and weight in relation to velocity may be included in pictures. This ensures that proper riprap sizes are selected to prevent scour at basin outlets. These large stones should be placed on a gravel protective bed to prevent piping of the finer bed material through the riprap.

## 5 Model Studies

General guides for design can be obtained from the preceding information with considerable accuracy. However, hydraulic model tests should be made on important, unusual spillways and energy dissipation devices. Model studies can provide design information and operational characteristics not easily visualized from mathematical calculations.

# New Words and Expressions

1.weir *n.* 低堰[拦河]坝，拦河[量水，溢流]堰
2.orifice *n.*（孔，管，出，开）口，注流孔，喷嘴[管，口]
3.uncontrolled crest spillway 无闸门的（无控制的）堰顶溢洪道
4.net length 净长度
5.pier *n.*（桥，墙，支，闸）墩
6.approach flow 行近水流
7.pier contraction coefficient 闸墩收束系数
8.velocity (energy) head 流速（能量）水头
9.ogee *a.*，*n.* S 形（的，曲线），双弯曲形（的）
10.rating *n.* 额定值，分等[类，配]，特性，规格
11.waterways experiment station 水道实验站
12.rounded nose 圆形墩头
13.partial *a.* 部分的，局部的，偏的，偏微分的 *n.* 偏导数
14.tainter gate 弧形闸门
15.approach channel 引水渠，引航道
16.flow pattern 流态
17.diversion dam 引水坝
18.orifice flow 孔口水流
19.dissipato *n.* 消能工，耗散[喷雾]器
20.energy dissipation 消能
21.hydraulic jump 水跃

22.kinetic *a.* （运）动的，活动的；~ energy 动能

23.plunge *v.*, *n.* 掉入，跳入，下（急）降

24.ski jump 滑雪式挑流（溢洪道）

25.frictional resistance 摩阻力

26.erosion resistance bed material 抗冲刷河床材料

27.diffusor 扩散器，扩散段，扩散式消能工

28.outlet control gate 泄水闸

29.stilling basin 静水池，消力池

30.momentum *n.* 动量，冲量

31.Froude number 弗汝德数

32.riprap *n.* 抛[堆，乱]石

# Lesson 16 Design Forces

The primary function of the dam is to raise the water level; therefore, the principal external force to be resisted by the dam is the pressure of impounded water. However, other forces also act on the structure. These forces (see Fig.2), which are discussed in the following sections, are: ①water pressure, external and internal, ②silt pressure, ③ice pressure, ④earthquake forces.

In a gravity dam the vertical weight of the dam is the main force resisting the water pressure. In a sloping face buttress dam part of the water load stabilizes the structure. In arch dams, the water load is transmitted through arch action into the foundation; weight becomes less of a stability component. Primary forces have certain valuations that must be considered in any design. The type of design modifies the transmission of forces and safety factors.

The dam must be stable enough to preclude overturning, sliding, overstressing, and any secondary erosion that might cause sliding of the foundation. The designer should consider design forces carefully to determine the type of dam needed and the forces that will be applied to the structure.

## 1 Water Pressure

The unit pressure of water increases in proportion to its depth. The water pressure acting normal to the dam face is represented by a triangular load distribution. The resultant of the load distribution is two-thirds of the distance from the water surface to the base of the section under consideration. The equation for unit water pressure is

$$p = w\,h$$

where: $w$ is unit weight of water (usually 62.5 lb. per cubic foot); $h$ is distance in feet from water surface to point of consideration.

The resultant water pressure is given by the equation

$$P_w = wh^2/2$$

In small gravity dams the upstream face is generally vertical; therefore, the water pressure is computed by this equation. As the height increases, a slightly inclined vertical face is generally used in design. The total vertical water load on such sections is represented by the weight of the volume of water vertically above that section. The resultant of the vertical water load acts through the centroid of that area. This stabilizing load is usually neglected for small dams.

In buttress dam design the water load stabilizes the face that is inclined several degrees from the vertical. In this case the water load is transmitted to the surface plane and the resultant increments of load must be computed for the various increments of depth.

The internal or uplift forces occur as pressures in the pores, cracks, and seams in both the dam and dam foundation. The void spaces within the concrete and the foundation material are filled with water that exerts a pressure in all directions. The uplift pressure intensity depends on the waterheads, that is ,the reservoir depth and the distance from the upstream face to the point under consideration. Uplift pressures occur in concrete and rock foundations as well as in soft pervious foundations. The total uplift force used in design is largely a matter of judgment based upon the character of the foundation, the steps taken to eliminate percolation, the probable deficiency of foundation drains, and the construction methods.

For hollow and buttress dams, the spaces between the buttresses relieve the uplift forces. However, when these dams are placed on soft pervious foundations, care must be exercised to avoid piping of the foundation material through the drain outlets.

Uplift pressures under a concrete dam on a pervious foundation are related to seepage through permeable materials. Water flowing through the pervious material is slowed down by frictional resistance. The amount and intensity of uplift flow under the dam must be considered where dams are constructed on pervious foundations. Uplift is important for all dams on pervious foundations.

The same methods of reducing uplift forces apply to all types of foundations. These methods include grouting a nearly impervious core wall at the upstream face of the dam, providing drains near the upstream face of the dam to allow water a free outlet, various cutoff walls, or combinations of safety measures.

The presence of seams, fissures in rock foundations, and the flow underneath the dam in pervious foundations all require certain assumptions for uplift forces. For a rock foundation, it is safe to assume a straight line variation from headwater to tailwater pressures as a measure of uplift. This pressure will be over the entire area of the dam. Any other uplift variation should be verified by electric analogue methods or comparative analysis to similar existing structures. The details of uplift forces for dams on pervious foundations must be determined from a flow net analysis that includes properly placed aprons, cutoffs, drainage, and other devices to control the intensity of uplift.

The uplift pressure at point A, is computed by the Westergaard equation

$$P_u=H_2+kx(H_1-H_2)/L$$

where the terms are defined in reference to Fig.16.1. In this equation, $P_u$ is in feet of water and is converted to unit pressure by multiplying by the term $w$, unit weight of water. The uplift factor $k$ refers to the location of the drainage system and its effectiveness in reducing uplift. Drains placed near the upstream face, behind a grout curtain, permit a reduction of $k$ from 1.0 (no drainage system) to 0.5. The reduced pressure varies over the total area and is assumed as a straight-line reduction shown by the dashed line in Fig. 16.2.

Foundation drainage on small dams is seldom economically feasible; total uplift is assumed in design. However, for moderate dam heights, the designer should consider an inspection gallery with proper relief drains in the concrete and in the foundation. Relief wells in the dam drainage gallery are placed vertically and on about 10-ft. centers. In the foundation, they are drilled from the gallery to a depth of four-to six-tenths the hydrostatic head or two-thirds the depth of the cutoff wall or grout curtain.

**(to be continued)**

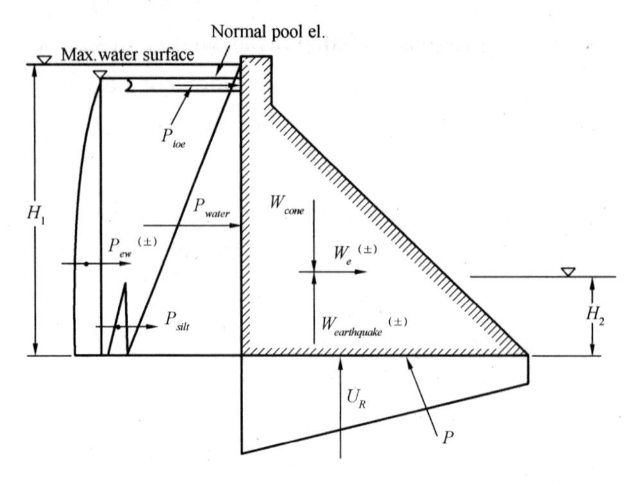


Fig. 16.1 Diagram of external forces resisted by a dam

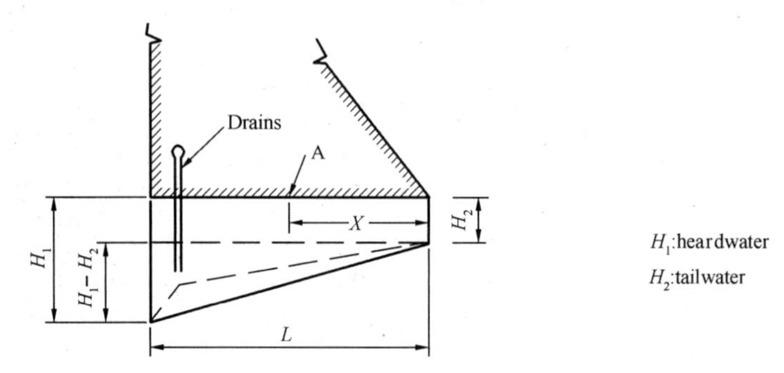


Fig. 16.2 Diagram of uplift pressures in the situation of the drainage system and non-drainage system

# New Words and Expressions

1.resultant *a.* 合成的，总计的；*n.* 合力，结果
2.centroid *n.*矩心；形心，重心
3.increment *n.*增加，增量
4.seam *n.*夹层，接缝，节理，裂缝
5.impervious curtain 不透水帷幕
6.headwater *n.* 库水，上游水
7.cutoff *n.* 截水墙，排水管，防渗幕
8.gallery *n.*长廊，美术馆，坑道
9.drainage gallery 排水廊道
10.hydrostatic head 静水头
11.grout *n.*水泥浆，砂浆；*v.* 灌浆
12.dashed line 虚线
13.centers 中心距
14.occur as 以…形式出现
15.care must be exercised 必须满意
16.on…centers 中心距为…
17.a matter of… 是个…问题
18.convert to 转化成

# Reading Material Forces

**(continue)**

## 2 Silt Pressure

Nearly all streams carry silt during normal and flood flows. Some silt flow is deposited in the reservoir created by the dam. If allowed to accumulate on the upstream face of the dam, the silt exerts loads greater than hydrostatic pressures. For small dams it is safe to assume that the silt load has a unit pressure equivalent to that of a fluid weighing 85 lb. per cubic foot and a vertical weight of 120 lb. per cubic foot. Sometimes the water-suspended silt is carried past the dam in special conduits, thus preventing deposit on the upstream face of the dam.

As the river flow is increasingly controlled, the silt load will be less important. Generally, a silt load develops slowly upon the dam face. As a result, the silt settlement tends to consolidate and partially support itself in the reservoir. For most small gravity and arch dams, the silt load is not usually important. However, on buttress dams with a sloping face, this accumulation may increase pressures significantly.

## 3 Ice Pressure

Ice pressures are produced by a thermal expansion in the ice sheet and by wind drag. When subjected to a temperature increase, ice expands and exerts a thrust against the upstream face of a dam Depending on the rate of temperature change, ice thickness, and other environmental conditions, a design ice pressure of 8 to 20 kips per linear foot is usually assumed.

Ice pressures are significant in all types of dam design. In gravity and buttress dams where overflow spillway and spillway gates are common, the gates should be heated to prevent ice from forming, Structural thickness at the dam crest must be sufficient to withstand stresses created by the ice sheet.

The force, Fig.16.3, exerted by expanding ice depends on ice thickness, rate of temperature rise in the ice, fluctuations on the water surface, reservoir shores, slope of upstream face of the dam, and wind drag.

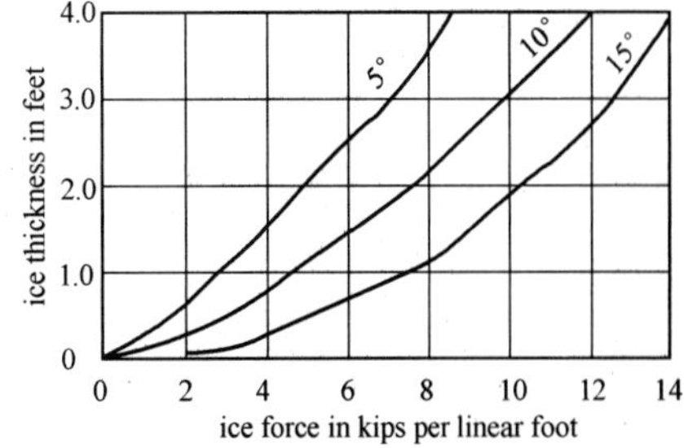

Fig. 16.3 Diagram of relation between ice pressures and ice thickenss

The rate of temperature rise in the ice is a function of the rate of rise of the air temperature and the snow cover on the ice. Lateral restraint of the ice sheet depends on the character of the reservoir shores and slope of the upstream face of the dam. Dams with slightly inclined slopes tend to resist ice pressure better than those with vertical slopes. Shores that are flat tend to hold the ice sheet from movement and this may lessen the force upon the dam. In small dams the ice problem is significant in the design of control structures, spillways, and gated devices.

## 4 Earthquake Forces

Earthquakes impart accelerations to the dam. Both vertical and horizontal earthquake loads are produced from these accelerations. To determine the total forces due to an earthquake, the intensity or acceleration due to the seismic motion must be established. Accelerations are expressed as percentages of gravity forces. In areas not subjected to extreme earthquake conditions, a horizontal acceleration of 0.1g and a vertical acceleration of 0.05g are generally used for design.

The force to accelerate the mass, $M$, of a dam is found from the equation

$$P_{ec} = Ma = \frac{W}{g}\alpha g = \alpha W$$

where: $P_{ec}$ is horizontal earthquake force; $a$ is earthquake acceleration; $g$ is acceleration of gravity; $W$ is weight of the dam or block; $\alpha$ is ration of $a$ to $g$.

The force $P_{ec}$ acts through the center of gravity of the section being studied.

The inertia force in pounds per square foot of the water is found by

$$P_{ew} = C\alpha wh$$

where: $C$ is dimensionless coefficient for the distribution and magnitude of pressures; $\alpha$ is ration of earthquake acceleration to acceleration of gravity, a/g; $w$ is unit weight of water, pounds per cubic foot; $h$ is total depth of reservoir water in feet; $y$ is vertical distance from reservoir surface to elevation under study in feet.

The dimensionless coefficient is defined in terms of the face slope and its maximum value $C_m$ by

$$C = \frac{C_m}{2}\left[\frac{y}{h}\left(2-\frac{y}{h}\right)+\sqrt{\frac{y}{h}\left(2-\frac{y}{h}\right)}\right]$$

Values of $C$ are obtained from Fig.16.4. The total horizontal force $V_e$, above any elevation y, distance below the reservoir surface, and the overturning moment, $M_e$, above that elevation are given $V_e = 0.726\ P_{ew}\ y$ and $M_e = 0.299\ P_{ew}\ y^2$.

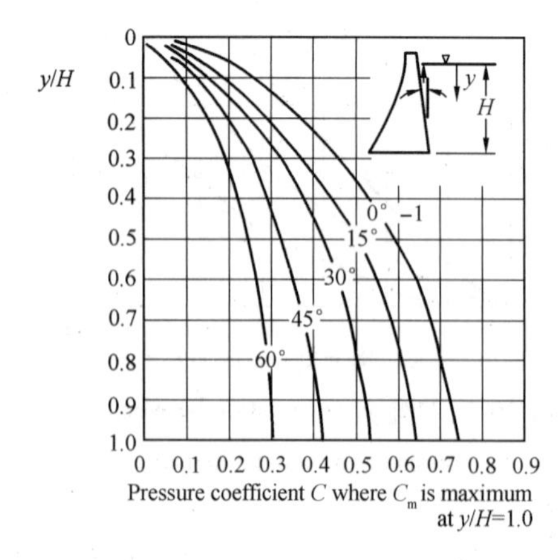


Fig. 16.4 Diagram of relation between pressure coefficient $C$ and $y/H$

Vibration or resonance caused by earthquakes is not likely to occur in low-to-moderate-height dams. Therefore, this element is not usually a design problem for low dams.

The effect of inertia on the concrete should be applied at the center of gravity of the mass regardless of the shape of the cross-section. Vertical motion may also occur during an earthquake with a resultant vertical inertia force that acts momentarily to reduce the effective weight of the dam. The water load tends to produce an overturning moment on the dam. The inertia force in the vertical motion upward tends to cause the concrete and water above the dam face to weigh less. This reduces the stability of the structure against sliding forces. Other forces that may be pertinent are wave pressures and wind loads.

## New Words and Expressions

1.water-suspended silt 水中悬浮泥沙
2.ice sheet 冰层
3.linear foot 延英尺，每英尺长
4.kips(=kilo-pounds) 千磅
5.fluctuation *n.* 波动，起伏，知降
6.reservoir shores 水库岸边
7.wind drag 风的曳引力
8.seismic motion 地震运动
9.inertia *n.* 惯性，惯量
10.dimensionless coefficient 无量纲系数
11.vibration *n.* 振动，摆动
12.resonance *n.* 共振（点，现象），轮调，中介
13.apt to 会，倾向于，易于

# Lesson 17 Significance of Infiltration and Process of Infiltration

## 1 Significance of Infiltration

Infiltration is the movement of water into the soil. There is a maximum rate at which the soil in a given condition can absorb water; this upper limit is called the infiltration capacity of the soil. If rainfall intensity is less than this capacity, the infiltration rate will be equal to the rainfall rate, whereas if rainfall intensity exceeds the ability of the soil to absorb moisture, infiltration occurs at the capacity rate. The excess of rainfall over infiltration collects on the soil surface and runs over the ground to streams. Infiltration rates are expressed in units of depth per unit time, the same as rainfall intensities. They refer to the depth of a sheet of water that would soak into the soil in a chosen time interval.

The soil surface is a filter that determines the path by which rainwater reaches a stream channel. Water that does not infiltrate runs quickly over the ground surface, whereas water entering the soil moves much more slowly underground. The soil, therefore, plays a major part in determining the volume of storm runoff, its timing, and its peak rate of flow. These are all of importance to the hydrologist interested in the planning of culverts, bridges, and other small structures. In running over the ground surfaces, water is capable of eroding topsoil and important organic residues on the land surface. The soil conservationist is concerned with either inducing this overland flow to infiltrate or conducting it safely away from fields or farm structures. Geomorphologists are also concerned with the magnitude, frequency, and spatial characteristics of infiltration relative to rainfall intensity, because overland flow is an important agent of landscape development.

The water that infiltrates the soil controls to some extent the water available for evapotranspiration. This supply of soil moisture is the effective rainfall as far as plants are concerned. Ecologists and agriculturists, therefore, need to refine their understanding of the relationships between plants and their water supply by considering infiltration and runoff. Infiltrated water that is not returned to the atmosphere by evapotranspiration reaches the groundwater system and supplies streamflow. Increasing the amount of infiltration may augment streamflow during dry weather, which is important for water supply, waste dilution, and other uses.

## 2 Process of Infiltration

Close examination of a lump of soil or the sides of a pit dug through a soil profile reveals that soils consist of millions of particles of sand, silt, and clay, separated by channels of different sizes. These channels include shrinkage cracks, wormholes, root holes, spaces between lumps or “crumbs” of soil and very fine spaces between the individual particles themselves. Such crack, holes, and fine spaces are called soil pores.

When rainfall reaches the ground surface, some or all of it enters soil pores. It is drawn into soil by the forces of gravity and capillary attraction. The rate of entry of water by free gravity flow is limited by the diameters of the pores. As water moves along such pores, it is subject to flow resistance, which increases as the diameter of the pore decreases. Under the influence of gravity, water moves vertically downward through the soil profile. On the other hand, capillary forces may move water vertically up or down or with a horizontal component. They act to draw water into the narrower pores, just as capillary forces drawing water up in a narrow glass tube are greater than those in a wide tube. Although such forces are strongest in soils with very fine pores, the pores may be so small that there is considerable resistance to flow through them. In large pones, such as wormholes or root holes, capillary forces are negligible and water moves downward under free gravity drainage. While flowing down through these passages, water is subject to lateral capillary forces, which draw it into the finer intergranular spaces leading off from the larger channel. Infiltration, therefore, involves three interdependent processes: entry through the soil surface, storage within the soil, and transmission through the soil.

## New Words and Expressions

1.infiltration *n.* 渗入，渗透
2.exceed *vt.* 超过
3.soak *v.* 浸，浸透
4.filter *v.* 过滤层，滤池
5.infiltrate *v.*渗入，吸入
6.culvert *n.* 涵洞，阴沟
7.organic *a.* 有机的
8.residue *n.*废料，沉淀
9.conservationist *n.*（自然资源）保护论者
10.induce *vt.* 导致
11.geomorphologist *n.* 地貌学家
12.agent *n.* 因素
13.landscape *n.* 地形，地貌
14.ecologist *n.* 生态学家
15.augment *v.* 增大，扩大
16.dilution *n.* 稀释，冲淡
17.pit *n.* 槽，洞，坑
18.profile *n.* （纵）断面
19.shrinkage *n.* 收缩；
shrinkage crack 收缩裂缝
20.wormhole *n.* 条虫状孔，虫孔
21.crumb *n.* 碎屑，屑粒
22.negligible *a.*微不足道的，不重要的
23.intergranular *a.*（颗）粒间的
24.entry *n.*进入，入口
25.transmission *n.*传递
26.the excess of…over… 超过…的部分；…比…多出部分
27.induce…to do sth. 促使（导致）…（做某事）
28.overland flow 表面径流，坡面流
29.a lump of 一块
30.to some extent 在某种程度上
31.as (so) far as…is concerned 就…而论
32.lead off 导出；引走，排除

## Reading Material Control of Infiltration

Many factors influence the shape of the infiltration capacity curve. But the most important variables are rainfall characteristics, soil properties, vegetation, and land use.

Long, intense rainfall packs down the loose soil surface, disperses fine soil particles, and causes them to plug soil pores. Rainstorms of long duration fill up the storage potential of the soil and cause clays to swell or may even saturate the soil completely.

Coarse-textured soils such as sands have large pores down which water can easily drain, while the exceedingly fine pores in clays retard drainage. If the soil particles are held together in aggregates by organic matter or a small amount of clay, the soil will have a loose, friable structure that will allow rapid infiltration and drainage. The depth of the soil profile and its initial moisture content is important determinants of how much infiltrating water can be stored in the soil before saturation is reached. Deep, well-drained, coarse-textured soils with a large content of organic matter, therefore, will tend to have high infiltration rates. And shallow soil profiles developed in clays will accept only low rates and volumes of infiltration. If the soil is invaded by dense ice lenses known a "concrete frost", its infiltration capacity may be reduced almost to zero. Although "porous frost" is formed by the growth of disseminated ice crystals, the soil does not reduce the rate of water intake.

Vegetative cover and therefore land use are very important controls of infiltration. Vegetation and litter protect soil from packing by raindrops and provide organic matter for binding soil particles together in open aggregates. Soil fauna that live on the organic matter assist this process by churning together the mineral particles and the organic material. The manipulation of vegetation during land use cause large differences in infiltration capacity under the same rainfall regime and sole type. In particular, the stripping of forests and their replacement by crops that do not cover the ground efficiently and do not maintain a high organic content in the soil often lower the infiltration capacity drastically. The problem is further aggravated by land management practices, such as plowing, which break up soil aggregate. Or if the soil surface is compacted by vehicles or trampled by livestock. The most extreme reduction of infiltration capacity, of course, involves the replacement of vegetation by an asphalt or concrete cover in urban areas. Each of these reductions of infiltration capacity increases the amount of surface runoff, and can produce soil erosion, flooding, and other economic and social costs, including the cost of practices such as soil conservation, stream channel modification, and other attempts to minimize damages caused by excessive runoff.

## New Words and Expressions

1.disperse *v.* 扩散，散开

2.coarse-textured *a.* 粗结构的

3.disseminate *vt.* 浸染；散布

4.aggregate *n.* 集合，集合体

5.churn *v.* 搅动，翻腾起

6.aggravate *vt.* 使恶化，使更严重

7.manipulation *n.* 使用，变换

8.asphalt *n.* 沥青，柏油

9.regime *n.* 情况，状态

# Lesson 18 Evapotranspiration

Evaporation is the transfer of water from the liquid to the vapor state. Transpiration is the process by which plants remove moisture from the soil and release it to the air as vapor. More than half of the precipitation which reaches the land surfaces of the earth is returned to the atmosphere by the combined processes, evapotranspiration. In arid regions evaporation may consume a large portion of the water stored in reservoirs.

The rate of evaporation from a water surface is proportional to the difference between the vapor pressure at the surface and the vapor pressure in the overlying air (Dalton's law). In still air, the vapor-pressure difference soon becomes small, and evaporation is limited by the rate of diffusion of vapor away from the water surface. Turbulence caused by wind and thermal convection transports the vapor from the surface layer and permits evaporation to continue.

Transpiration is essentially the evaporation of water from the leaves of plants. Rates of transpiration will, therefore, be about the same as rates of evaporation from a free water surface if the supply of water to the plant is not limited. Estimated free water evaporation may, therefore, be assumed to indicate the potential evapotranspiration from a vegetated soil surface.

The total quantity of transpiration by plants over a long period of time is limited primarily by the availability of water. In areas of abundant rainfall well distributed through the year, all plants will transpire at about the same rates and the differences in total will result from the differences in the length of the growing seasons for the various species. Where water supply is limited and seasonal, depth of roots becomes very important. Here, shallow-rooted grasses wilt and die when the surface soil becomes dry while deep-rooted trees and plants will continue to withdraw water from lower soil layers. The deeper-rooted vegetation will transpire a greater amount of water in the course of a year. The rate of transpiration is not materially reduced by decreases in soil moisture until the wilting point of the soil is reached.

Evapotrspiration, sometimes called consumptive use or total evaporation describes the total water removed from an area by transpiration and by evaporation from soil, snow, and water surfaces. An estimate of the actual evapotranspiration from an area can be made by subtracting measured outflow from the area (surface and subsurface) from the total water supply (precipitation, surface and subsurface inflow, and imported water). Change in surface and underground storage must be included when significant.

Several attempts have been made to relate evapotranspiration to climatological data through simple equations such as

$$U_c=0.9+0.00015\Sigma(T_{\max}-32) \tag{18.1}$$

where: $U_c$ is the consumptive use in feet and $\Sigma(T_{\max}-32)$ is the sum of the growing season maximum temperatures less 32℉. With $U_c$ in centimeters and temperatures in degrees Celsius,

Eq. (18.1) becomes

$$U_c=27.4+0.00823\Sigma T_{\max} \tag{18.2}$$

Such formulas agree fairly well with average values of annual evapotranspiration over a period of years, but it is clear that the evaporative process is too complex to be well defined by a simple temperature function.

As indicated before, the potential evapotranspiration from an area can be estimated from the free water evaporation. Actual evapotranspiration equals the potential value $E_{pot}$ as limited by the available moisture. On a natural watershed with many vegetal species, it is reasonable to assume that evapotranspiration rates do vary with soil moisture since shallow-rooted species will cease to transpire before deeper-rooted species. A moisture-accounting procedure can be established using the continuity equation

$$P-R-G_0-E_{act}=\Delta M \tag{18.3}$$

where: $P$ is precipitation; $R$ is surface runoff; $G_0$ is subsurface outflow; $E_{act}$ is actual evapotranspiration; $\Delta M$ is the change in moisture storage.

$E_{\text{act}}$ is estimated as

$$E_{act} = E_{pot}\ M_{act}/\ M_{\max} \tag{18.4}$$

where: $M_{act}$ is the computed soil moisture storage on any data and $M_{\max}$ is an assumed maximum soil moisture content. A moisture-accounting procedure of this type may be used to calculate runoff as well as to estimate evapotranspiration.

## New Words and Expressions

1.evapotranspirtion *n.* 腾发（量）
2. be (directly/inversely)…to 与…成正/反比（例）
3.turbulence *n.* 扰动，紊（湍）流性
4.wilt *n.*，*v.* 凋萎（残）
5.materially *ad.* 实质、实际上，显著地
6.consumptive *a.* 消费的，消耗的
7.climatological *a.* 气候的
8.Celsius *a.*，*n.*摄氏（温度）
9.Fahrenheit *n.*，*a.* 华氏（温度）
10.vegetal *a.* 植物的；*n.* 植物
11.Dalton's low 道尔顿定律

## Reading Material Evaporation

Evaporation is the name given to the change of water from the liquid state to the gaseous state. Exactly how much water is evaporated is very difficult to judge accurately, although water vapor on the average is about 2% of the total volume of the atmosphere. Recent studies have measured the evaporation from different sized bodies of water at a constant temperature, over a given period of time. The findings of these studies will be incorporated in the following sections.

### 1 Sources of Water Vapor

Water vapor in the atmosphere has been evaporated from various sources. The oceans, covering approximately three-quarters of the earth's surface, provide most of the water vapor. Other sources are of relatively minor importance. They include rivers, creeks and inland water storage, which are continuously providing vapor for the atmosphere, as is the moist land surface itself, particularly in tropical areas. Plants also give off moisture, a process referred to as transpiration.

Once the atmosphere carrying this moisture moves over land areas and cools, the vapor is converted back into the liquid or solid state and, on reaching the earth, either is evaporated again or travels over the surface according to gravity.

## 2 Factors Affecting Evaporation

As evaporation is by no means constant, any attempt to relate total amounts directly to one or two factors is futile. The only generalizations which can be made with any degree of certainty are that the water body must be heated and favorable conditions must exist in the atmosphere. Some of the main factors retarding or promoting evaporation are listed below.

### 2.1 Vapour Pressure

There is a direct relationship between evaporation and water temperature. As the temperature increases, the vapor pressure of the water increases rapidly. Vapor pressure means the ability of water molecules to break away from the mass and "fly" into the air. In other words, when the vapor pressure of the overlying air is below the vapor pressure of the water surface, water molecules will diffuse into the air mass. The general law has perhaps best been stated by Harding:

"If other factors remain constant, evaporation is proportional to the deficit in vapor pressure, which is the difference between the pressure of saturated vapor at the temperature of the water and the vapor pressure of the air."

### 2.2 Air Temperature

The temperature of the overlying atmosphere determines the rate of vapor removal, while, as stated above, the water temperature determines the rate of escape of water molecules into the atmosphere. With a dry, stationary air mass over a water surface, evaporation will continue until the air becomes saturated. If the air temperature is constant, there will be a limit to the evaporation. On the other hand, if the overlying mass is of the same temperature but saturated, that is ,holding its entire capacity of water vapor, the water molecules being diffused from the water surface will return to it, as the vapor pressures are the same. With an increase in temperature, evaporation will recommence, as the air is now capable of holding more water vapor.

### 2.3 Relative Humidity

This is the amount of water vapor in the air in relation to the amount that can be held in the air at that temperature, or the quantity held in relation to the capacity at that temperature. This ratio is always expressed as a percentage. Temperature change has an inverse effect on reduction or increase of the relative humidity and the capacity of the air. If the temperature of the air is 20℃ and the air is holding 80% of its vapor capacity, then when the temperature rises, the relative humidity falls. Conversely, if the temperature falls, the relative humidity rises.

### 2.4 Wind

Winds play an important part in the evaporation process as they enable the moist air over the water body to be replaced by drier air. If the moist air were not replaced, the stationary air would become saturated and unable to hold further water vapor.

Once wind strength reaches a certain velocity any further increase will not increase the rate of evaporation. If the air speed is slightly greater than the evaporation rate (in other words, is capable of removing all the moist air) then further increases will have no effect. This limiting velocity varies with atmospheric temperature and pressure.

The movement of air may be critical. A highly turbulent air mass with marked vertical currents will distribute water vapor throughout the entire air mass, thus reducing the load at lower levels. It will therefore be more capable of removing water vapor from just above the water surface as the particles are diffused by the water temperature. A laminar flowing air mass, in which layers of air molecules pass over one another with no vertical redistribution, will simply replace moist air by further moist air, as the horizontal movement does not introduce drier air streams from higher altitudes.

### 2.5 Barometric Pressure

If all other factors remained constant, evaporation will increase with a decrease of barometric pressure. Theoretically, evaporation increases with increasing altitude, as the barometric pressure is lower. Normally, however, the decrease in temperature, and other responses typical of increasing altitude, counteract reduction of pressure, and so evaporation is less.

### 2.6 Nature of the Water

Evaporation varies inversely with the degree of salinity of the water until the water has a salty content of about 30%. At this stage, in calm water, salt crusts tend to appear and these restrict further evaporation. Turbulence may allow evaporation to continue by freeing the water of its crust.

Generally, for every 1% increase in salinity, evaporation decrease by 1%. Fresh water then, per unit area, has a higher rate of evaporation than salt water. Over the seas, evaporation may be decreased by about 3%~5% in comparison with fresh water, given the same volume, depth and temperature.

### 2.7 Storage Variations

As evaporation is ultimately determined either directly or indirectly by insolation, it is reasonable to expect that the great oceans will evaporate less water per unit of time than inland water storage. The reasons for this are twofold: ①the salinity content discussed under point 6 above, ② the sizes of the water mass.

If two bodies of water have the same surface area exposed to the sun's rays, the deeper one will have less evaporation in a given time. This is because the insolation is dispersed through a greater area and so it takes longer for the mass to reach the same temperature as the shallow water body. In spring and early summer the deep storage will absorb heat which would otherwise be used for evaporation. During autumn and winter the heat stored in the water mass is released, adding to the heat present in the atmosphere. Thus evaporation is relatively low during spring and summer, and relatively high during autumn and winter for deep storage.

After intensive research by the United States Weather Bureau it was found that evaporation from a tank measuring 1.2 meters in diameter and 25 centimeters deep (termed a Class A pan), was greater that from a large, open water body. The latter rate was determined by using the reduction coefficient of 0.7.

## New Words and Expressions

1.finding *n.* （常用 *pl.*）调查（研究）的结果
2.minor *a.* 较次要的；较小的，较少的
3.creek *n.* 小溪，小湾，小河
4.transpiration *n.* 蒸腾，散发，排出
5.futile *a.* 无益的，无用的
6.generalization *n.* 概念，概况
7.certainty *n.* 确实，肯定，必然的事
8.retard *vt.* 阻止,阻碍，延迟； vi. 减速
9.promote *vt.* 促进；发扬；助长
10.overlie *vt.* 压在…上面，覆盖 现在分词是 overlying
11.diffuse *vt.* 扩散，散开；渗出
12.deficit *n.* 差值，赤字
13.saturated *a.* 饱和的，浸透的
14.removal *n.* 移动；调动
15.re-commence *vt.*，vi. （使）重新开始
16.humidity *n.* 温度，湿气，水分含量
17.inverse *a.* 相反的，倒转的
18.turbulent *a.* 紊动的，湍急的
19.laminar *a.* 层流的，层状的
20.barometric *a.* 气压（计）的，测定气压的
21.counteract *vt.* 抵消，抵抗，抵制，阻碍
22.graphically *ad.* 图解地
23.inversely *ad.* 相反地
24.salinity *n.* 含盐量，盐浓度，咸度
25.calm *a.* （天气，海洋等）静的，平静的
26.restrict *vt.* 限制，限定
27.insolation *n.* 日照，曝晒
28.two-fold *a.* , *ad.* 双重，两倍
29.disperse *vt.* 使疏开，使分散；vi. 分散，散开，散去
30.intensive *a.* 加强的,集中的，深入细微的
31.centimetre *n.* 厘米
32.free of 无…的；摆脱了…的
33.Class A pan A 级蒸发皿
34.reduction coefficient 折减系数
35.Harding 哈定
36.the United State Weather Bureau 美国气象局

# Lesson 19 Irrigation Methods

Irrigation methods can be divided into four main types-surface, subsurface, sprinkler, and trickle irrigation and many subtypes. Surface irrigation is the oldest type and still accounts for about three fourths of all irrigation. Subsurface irrigation is limited in its adaptation. Sprinkler irrigation can be used in any climate, is the most popular method in humid regions, and is still expanding in use. Trickle irrigation, the newest type, makes the most efficient use of water.

## 1 Surface Irrigation

Surface irrigation includes both furrow and flood types. Furrow irrigation is used with row crops by running water in the cultivated channel between the rows. The rows can be fed with siphon tubes or with spiles or in groups from ditch turnouts.

### 1.1 Furrow Irrigation

Furrow irrigation creates a serious erosion hazard because the water flows in the unprotected area between the rows where the soil has been loosened by cultivation. The maximum non-erosive stream flow in liters per second can be estimated from the equation:

$$Q_{\max} = \frac{0.6(\mathrm{L/s})}{i}$$

This equation should be used only for slopes of more than 0.3% because the flow on flatter slopes is usually limited by the furrow capacity rather than by erosion. The 0.6 factor should be decreased if the soil is known to be more erodible than average.

Irrigation furrows with slope gradients steeper than 2% are difficult to irrigate without erosion. Large streams would erode the soil and small streams will only flow a short distance before all the water infiltrates. Contour furrows overcome this problem because they are placed on a gradient of about 0.5% across the main slope. Extra care must be taken to be sure that water does not cut through any of the ridges between contour furrows. The resulting overload on the lower furrow could cause it to overtop and begin a chain reaction that would likely produce a gully.

The length of irrigation furrows is limited by the distance irrigation streams will flow during an irrigation period. Uniform irrigation requires that this period of time be no longer than one fourth of the total irrigation period. For example, the irrigation water should reach the end of the rows in two hours out of an eight-hour irrigation period or three hours out of a twelve-hour period. The maximum length of irrigation rows therefore depends on the infiltration rate and erodibility of the soil, the slope, and the amount (depth) of water to be applied. Table 19.1 contains estimates of the appropriate lengths of rows for various conditions.

Irrigation according to the principles outlined in the preceding paragraphs can achieve about 60% efficiency in the application of water. The other 40% is lost by evaporation, deep percolation in the upper ends of the rows and in the most permeable soil, and in wastewater from the lower ends of the rows. The wastewater loss can be lessened if the irrigator will reduce the size of the irrigation streams when they approach the lower end of the rows. The extra water can then be used elsewhere, perhaps on a pasture that needs a short irrigation period.

Table 19.1 Suggested maximum lengths in meters of cultivated furrows for different soils, slopes, and depths of water in cm to be applied

| Furrow slope (%) | Average depth of water applied (cm) | | | | | | | | | | | |
|---|---|---|---|---|---|---|---|---|---|---|---|---|
| | Clays | | | | Loams | | | | Sands | | | |
| | 7.5 | 15 | 22.5 | 30 | 5 | 10 | 15 | 20 | 5 | 7.5 | 10 | 12.5 |
| 0.05 | 300 | 400 | 400 | 400 | 120 | 270 | 400 | 400 | 60 | 90 | 150 | 190 |
| 0.1 | 340 | 440 | 470 | 500 | 180 | 340 | 440 | 470 | 90 | 120 | 190 | 220 |
| 0.2 | 370 | 470 | 530 | 620 | 220 | 370 | 470 | 530 | 120 | 190 | 250 | 300 |
| 0.3 | 400 | 500 | 620 | 800 | 280 | 400 | 500 | 600 | 150 | 220 | 280 | 400 |
| 0.5 | 400 | 500 | 560 | 750 | 280 | 370 | 470 | 530 | 120 | 190 | 250 | 300 |
| 1.0 | 280 | 400 | 500 | 600 | 250 | 300 | 370 | 470 | 90 | 150 | 220 | 250 |
| 1.5 | 250 | 340 | 430 | 500 | 220 | 280 | 340 | 400 | 80 | 120 | 190 | 220 |
| 2.0 | 220 | 270 | 340 | 400 | 180 | 250 | 300 | 340 | 60 | 90 | 150 | 190 |

Principles similar to those outlined for furrow irrigation apply to related methods of irrigation. Small furrows known as corrugations are used for uncultivated grain and forage crops. Such vegetation protects the soil better than row crops, but corrugations are too small to carry large streams of water. The row lengths are therefore similar to those for furrow irrigation.

### 1.2 Flood Irrigation

The three main types of flood irrigation are basin irrigation, border irrigation, and wild flooding. Basin irrigation is probably the oldest method of all. It was practiced in Egypt more than 5000 years ago. It is a simple method that is still widely used to keep land flooded for long periods for paddy rice production or for shorter periods for many other crops.

Land preparation for basin irrigation is accomplished by forming a narrow ridge between 15 and 50 cm high on all sides of each area to be flooded. The elevations within any one basin should be as uniform as possible—certainly within a range of 5 or 10 cm. The area of a basin may be limited by elevation changes, by the size that can be covered uniformly by the available water supply on permeable soils, or by cropping factors. Basins range in size from those designed to irrigate individual trees or small areas of vegetable crop to rice paddies occupying several hectares.

A ditch or other water supply large enough to flood the basin must be available on one side. Water is turned in until the desired depth is reached, then cut back to just enough to hold constant depth of about 10 cm for paddy rice or shut off completely for other crops. The water in the basin may be allowed to completely infiltrate or, in some low permeability soils, the excess may be drained onto a lower basin after a specified time.

Border irrigation can be described as elongated basins with a gentle slope in the long direction. Water is turned in at the upper end of the border and allowed to flow down its length as though it were a very wide furrow. Borders range from 3 to 30 m wide and must be nearly level across their width so the entire area will be irrigated uniformly. Their lengths are similar to the lengths of furrows on comparable soils and slope gradients shown in Table 19.1.

Border irrigation can be used with slope gradients between 0.2% and 2% for cultivated crops, up to 4% or 5% for small grain or hay crops, and up to about 8% for pastures. Extensive land leveling is often required because the topography must be smoother than for furrow irrigation. The cost of land leveling is offset by the low labor requirement for turning water into a few borders rather than into many furrows or corrugations. The smooth topography is easy to work across at harvest time.

Several variations of border irrigation have been devised. Irrigation terraces made with surfaces that either slope like borders or are level like basins are one example. In another variation, ditches replace the ridges between borders and the area between is irrigated by blocking the ditches so they flood the border. Unfortunately, erosion can be a problem in the ditches, and uniform water application is often difficult to attain with this method.

Wild flooding is used to irrigate forage crops and sometimes small grains on uneven topography. Water flows down the ridges in ditches and is diverted to flood across the land. It is often necessary to have small spreader ditches to redistribute water that naturally accumulates in swales. The irrigator uses a shovel to make small furrows and ridges to guide water to any areas that would otherwise remain dry.

Wild flooding is inefficient in use of water and labor, but it irrigates land that cannot be managed by other methods of surface irrigation. The soil may be too shallow or stony to have its surface smoothed by land leveling, and it may not be used intensely enough to justify a large investment.

Rolling topography with slope gradients up to about 10% can be irrigated by wild flooding with permanent close-growing vegetation.

## 2 Subsurface Irrigation

Subsurface irrigation, also called sub-irrigation, can be considered as a controlled drainage system. Ditches are usually used, but some systems use tile lines. The systems remove water during wet seasons and add it during dry seasons so the water table is always at a controlled depth. That depth might be as little as 30 cm for shallow-rooted vegetation in a coarse sandy soil or as great as 120 cm in some loamy soils. The surface soil should be dry but most of the root zone should be moist. The field can even be cultivated and irrigated at the same time.

The required conditions for subsurface irrigation are so stringent that relatively little land is sub-irrigated. The land surface must be quite smooth and have a slope gradient of less than 0.5%. The subsoil must be highly permeable, but it must have a shallow water table or be underlain by an impermeable layer that permits a perched water table to be maintained. Both the soil and the irrigation water must be low in salts to avoid the formation of saline and sodic soil. Suitable conditions for subsurface irrigation most often occur on glacial outwash plains, terraces, or deltas in humid or sub-humid areas.

## 3 Sprinkler Irrigation

Large-scale sprinkler irrigation is much newer than surface and subsurface irrigation because the necessary pipes, pumps, and power supply were not available until comparatively recent times.

Advantages such as portability, adaptability to a wide range of soil and topographic conditions with little or no land preparation, and good control of water application have made sprinkler irrigation popular. High efficiency of water application may result in energy savings and help avoid erosion and leaching nitrates from the soil. Disadvantages limiting its use include high equipment and operating costs, the need to move lines in muddy conditions, salt damage to some plants if poor-quality water is used, and disease problems with some plants.

Most field sprinklers use a rotating sprinkler head of the general type. Although the sprinklers may be fixed in permanent locations for limited areas of high-value crops, they are usually mounted on either moving or movable lines.

## 4 Trickle Irrigation

Trickle irrigation, also called drip irrigation, is the newest method and the one that achieves the highest irrigation efficiency: about 90% of the applied water is available to the plants. High efficiency is achieved by supplying water to individual plants through small plastic lines. Water is supplied either continuously or so frequently that the plant roots grow in constantly moist soil.

Trickle irrigation is especially suitable for watering trees or other large plants. Much of its use has been in orchards and vineyards but it has also been used to irrigate a variety of row crops including several kinds of vegetables and fruits. Its advantages are greatest where areas between plants can be left dry. It has no advantage for close-growing vegetation such as lawns, pastures, or small grain crops.

An Israeli engineer named Symcha Blass developed the idea of trickle irrigation in the 1930s, but it had to wait until plastic tubing was available to make a practical system. Trickle irrigation in the United States increased from 40 ha in 1960 to over 50,000 ha in 1976 out of a worldwide total of about 160,000 ha. Nearly half of the trickle irrigation in the United States is in California, some of it in avocado orchards with slopes up to 50% or 60%. Erosion is not a problem because there is no runoff.

A bonus with trickle irrigation is its ability to use water with a higher salt content than any other method up to about 2500mg/liter. The constant flow of water from the trickle emitter toward the outer edges of the plant root zone carries the salt along with it. Salt concentrations become very high in the dry areas between plants but not in the actual root zone.

Trickle irrigation saves water, is able to use water high in salt, functions well in all but the extremes of coarse and fine textured soils, works on almost any topography without causing erosion, and required little labor. The disadvantages are mainly high equipment costs and plugging of the lines by sediment, salt encrustation, or algae.

A trickle irrigation system normally includes a control box that regulates the water pressure, filters the water, and provides for the addition of fertilizers and herbicides. Chlorine may be added to eliminate algal growth. The water pressure for trickle irrigation is normally 0.4 to 1kg/cm$^2$ as compared to 1 to 8kg/cm$^2$ for sprinkler irrigation. Some trickle controls are set to increase the pressure periodically and flush the lines to reduce clogging.

Trickle irrigation lines branch into several parts at three or four stages to provide the many outlets required. The last stage is a flexible plastic lateral line 12 to 32 mm in diameter that lies either on or just below the soil surface and applies the water either through small holes in the line or through emitter nozzles. Emitter nozzles lead the water through a long spiral path that slows the flow and permits a larger emission hole to be used. The larger hole is less subject to plugging.

## New Words and Expressions

1.sprinkler *n.* 喷洒；人工降雨
2.trickle *n.* 滴（流）；细流
3.subtype *n.* 分（副，辅助）型
4.popular *a.* 通用的，受欢迎的
5.cultivate *vt.* 耕种；开垦；种植
6.spile *n.* 插管；小塞子；木塞
7.turnout *n.*田间取水口；渠道分叉口
8.unprotected *a.* 无防护设备的,未加保护的
9.loosen *vt.* 松散；分散
10.nonerosive *a.* 不浸蚀的
11.flatter *a.* 较平坦的
12.erodible *a.* 易受腐蚀，浸蚀，冲刷
13.steep *a.* 陡急的，峻峭的
14.ridge *n.* 畦，垄，埂
15.overtop *vt.* 超过
16.gully *n.* （冲刷，集水，排水）沟；沟渠；溪谷
17.erodibility *n.* 浸蚀；冲刷
18.preceding *a.* 以前的，前面的；上述的
19.lessen *vt.* 减（变）少；缩小
20.pasture *n.* 牧草（场）；*vt.*放牧
21.corrugation *n.* 田间浅沟；波状沟
22.uncultivate *vt.* 未耕种
23.paddy *n.* 水稻；谷；稻田
24.preparation *n.* 处理；加工
25.uniformly *ad.* 均匀地，无变化地
26.elongated *a.* 细长的，延长的
27.level *a.* 相等的；同程度的
28.hay *n.* 干（粮，牧）草
29.offset *n.* 弥补；抵消
30.terrace *n.* 梯田，阶地
31.uneven *a.* 不平的，不均的，不稳定的
32.spreader *n.* 喷洒器
33.swale *n.* 低洼地
34.shovel *n.* 铁铲
35.stony *a.* 石（头、质）的，多石的
36.intensely *ad.* 强烈地；一心一意地
37.close-growing *a.* 密集生长的
38.tile *n.* 暗管；瓦管
39.loamy *a.* 壤土的
40.stringent *a.* 格的；精确的
41.subsoil *n.* 底土；下层土
42.underlie (underlay, underlain) *vt.* 位于…下面
43.perched *a.* 位置高的
44.saline *a.* 含盐的
45.sodic *a.* 钠质的
46.outwash *n.* 冲刷；清除
47.delta *n.* 三角洲
48.subhumid *a.* 半湿润的
49.portability *n.* 轻便（性）；能够动
50.topographic *a.* 地形的
51.center-pivot *a.* 自转的
52.orchard *n.* 果树园
53.vineyard *n.* 葡萄园
54.lawn *n.* 草坪
55.avocado *n.* 鳄梨
56.bonus *n.* 附带的优点（好处）
57.emitter *n.* 滴头；放射机
58.fine-textured *a.* 细（粒）组织结构的

59.coarse-textured *a.* 粗（粒）组织结构的
60.encrustation *n.* 结晶；结壳
61.alga *n.* (*pl.*algae) 海藻；藻（类）
62.filter *vt.* 过滤；滤清
63.herbicide *n.* 除莠（草）剂
64.chlorine *n.* 氯（气）
65.algal *a.* 藻类的
66.clogging *n.* 阻（堵）塞；闭合
67.forage crops 饲料作物
68.flood irrigation 淹灌
69.basin irrigation 格田淹没，畦田漫灌
70.land preparation 田间准备工作
71.cut back 减少；降低
72.border irrigation 畦灌（法）
73.wild flooding 漫灌
74.subsurface irrigation = sub-irrigation 地下灌溉
75.tile lines 暗管，瓦管
76.trickle irrigation = drip irrigation 滴灌
77.trickle emitter 滴头
78.Symcha Blass 西姆查·布莱斯

## Reading Material Irrigation Methods

There are five basic methods of applying irrigation water to fields—flooding, furrow irrigation, sprinkling, sub-irrigation, and trickle irrigation. Numerous subclasses exist within these basic methods. Wild flooding consists in turning the water into natural slopes without much control or prior preparation. It is usually wasteful of water, and unless the land is naturally smooth, the resulting irrigation will be quite uneven. Wild flooding is used mainly for pastures and fields of native hay on steep slopes where abundant water is available and crop values do not warrant more expensive preparations. Controlled flooding may be accomplished from field ditches or by use of borders, checks, or basins. Flooding from field ditches is often adaptable to lands with topography too irregular for other flooding methods. It is relatively inexpensive because it requires a minimum of preparation. Water is brought to the field in permanent ditches and distributed across the field in smaller ditches spaced to conform to the topography, soil, and rate of flow. Under ideal conditions, the ditch spacing and flow rate should be such that the water will just infiltrate in the time it is flowing across the field. If the flow is too rapid, some of the water will not have time to infiltrate and surface waste will occur at the lower edge of the field. If flow is too slow, excessive percolation will occur near the ditch, and too little water will reach the lower end of the field.

The border method of flooding requires that the land be divided into strips 30 to 60 ft (10 to 20 m) wide and 300 to 1200 ft (100 to 400 m) long. The strips are separated by low levees, or borders. Water is turned into each strip through a head gate along one of the narrow sides and flows downhill the length of the strip. Preparation of land for border-strip irrigation is more expensive than for ordinary flooding, but this may be offset by a decrease in water waste because of the improved control. Check flooding is accomplished by turning water into relatively level plots, or checks, surrounded by levees. If the land is initially level, the plots may be rectangular but with some initial slope the checks will usually follow the contours. Check flooding is accomplished by turning water into relatively level plots, or checks, surrounded by levees. If the land is initially level, the plots may be rectangular but with some initial slope the checks will usually follow the contours. Check flooding is useful in very permeable soils where excessive percolation might occur near a supply ditch. It is also advantageous in heavy soils where infiltration would be inadequate in the time required for the flow to cross the field. In check flooding the check is filled with water at a fairly high rate and allowed to stand until the water infiltrates. The basin-flooding method is check flooding adapted to orchards. Basins are constructed around one or more trees depending on topography, and the flow is turned into the basin to stand until it infiltrates. Portable pipes or large hoses are often used in place of ditches for conveying water to the basins.

Furrow irrigation is widely used for row crops. The furrow is a narrow ditch between rows of plants. An important advantage of the furrow method is that only 0.2 to 0.5 as much surface area is wetted during irrigation as compared with flooding, and evaporation losses are correspondingly reduced. Furrow irrigation is adapted to lands of irregular topography. Customarily the furrow is run normal to the contours, although this should be avoided on steep slopes where soil erosion may be severe. Spacing of furrows is determined by the proper spacing of the plants. Furrows vary from 3 to 12 in (10 to 30 cm) deep and may be as much as 1500 ft (500 m) long. Excessively long furrows may result in too much percolation near the upper end and too little water at the down slope end. Water may be diverted by an opening in the bank of the supply ditch, but many farmers now use small siphons made out of 4 ft lengths of plastic tubing about 2 in, in diameter. These siphons are easily primed by immersion in the ditch and provide a uniform flow to the furrow without the necessity of damaging the ditch bank.

The development of lightweight pipe with quick couplers resulted in a rapid increase in sprinkler irrigation after World War II. Sprinkler irrigation offers a means of irrigating areas which are so irregular that they prevent use of any surface-irrigation methods. By using a low supply rate, deep percolation or surface runoff and erosion can be minimized. Offsetting these advantages is the relatively high cost of the sprinkling equipment and the permanent installations necessary to supply water to the sprinkler lines. Very low delivery rates may also result in fairly high evaporation from the spray and the wetted vegetation. In recent years, high labor costs for surface irrigation have increased the attractiveness of sprinkler irrigation. During the period from 1958 to 1967 the total irrigated acreage in the United States changed from 36 to 45 million acres (14.4 to 18 million hectares) an increase of 25 percent , while sprinkler use rose from 3.7 to 7.6 million acres (1.5 to 3.1 million hectares), a 106 percent increase. Sprinkling may be accomplished with fixed perforated pipe, rotation sprinkler heads, or fixed sprinkler heads. It is impossible to get completely uniform distribution of water around a sprinkler head, and spacing of the heads must be planned to overlap spray areas so that distribution is essentially uniform.

In a few areas soil conditions are favorable to sub-irrigation. The required conditions are a permeable soil in the root zone, underlain by an impermeable horizon or a high water table. Water is delivered to the field in ditches spaced 50 to 100 ft (15 to 30 m) apart and is allowed to seep into the ground to maintain the water table at a height such that water from the capillary fringe is available to the crops. Low flow rates are necessary in the supply ditches, and free drainage of water must be permitted, either naturally or with drainage works, to prevent waterlogging of the fields. The irrigation water should be of good quality to avoid excessive soil salinity. Sub-irrigation results in a minimum of evaporation loss and surface waste and requires little field preparation and labor.

In trickle (or drip) irrigation a perforated plastic pipe is laid along the ground at the base of a row of plants. The perforations are designed to emit a trickle (5 liters/hr or less) and spaced to produce a wetted strip along the crop row or a wetted bulb at each plant. The main advantage of trickle irrigation is the excellent control, since water can be applied at a rate close to the rate of consumption by the plant. Evaporation from the soil surface is minimal and deep percolation almost entirely avoided. Nutrients can be applied directly to the plant roots by adding liquid fertilizer to the water. Salinity problems are minimal for the salts move to the outer edge of the wetted zone away from the roots. Investment costs are high, but labor costs are low once the system is set up. The technique of trickle irrigation was developed in Israel.

## New Words and Expressions

1.pasture *n.* 牧场；*v.* 放牧
2.hay *n.* 干草；*v.* 种草（供制干草田）
3.warrant *n.*（正当）理由，保证；*vt.*担保，保证，批准
4.check *v.*, *n.* 校对，阻止，支票 *pl.*格田
5.adaptable *a.* 适合的，适用于
6.be-to 适用于…的
7.conform *vt.* 使一致，使遵照；*vi.* ~to/with 符合，相似
8.strip *v.* 剥去，拆开；*n.*带，长条，行
9.offset *n.*，*vt.* 偏移，失调，补偿
10.be-by M 为M所抵销（补偿）
11.hase *n.* 长统袜，软管；*vt.* 用水管浇（洗）
12.convery *vt.* 运送，搬运，传输
13.customary *a.* 通常，惯例的；
customarily *ad.*
14.divert *v.* 转移，（使）转向，（使）高兴，分出
15.immerse *vt.* 沉浸，使陷入；-sion *n.*
16.coupler *n.* 连结着，连接器，接火头
17.delivery *n.* 交付（货），投递，传送
18.acreage *n.* 英亩数，土地面积
19.perforate *v.* 穿孔；*a.* 有孔的（排孔）
20.overlap *v.* 交搭，重迭
21.underlay *vt.* (underlaid)垫，衬；*n.* 衬垫物
22.fringe *n.* 边缘
23.bulb *n.* 球形的
24.nutrient *a.* 营养的；*n.* 营养物
25.head gate 渠首闸；
head gates 船闸上闸门
26.normal to 垂直于
27.make out of 用…制造出，理解，离开
28.spray area 喷洒面积
29.drainage works 排水系统

# Lesson 20 Irrigation Canals

The problem of canal location is similar in many respects to highway location, but the solution may be more difficult since the slope of the canal bottom must be downgrade, and frequent changes in slope (and hence changes in section) should be avoided. Within the limitations of topography, the exact route of a canal is determined by the slopes that can be tolerated. Excessive slope may result in a velocity sufficient to cause erosion of the channel bottom or sides. The velocity at which scour will begin depends on the bed material and the shape of the channel cross section. Fine-grained soils generally scour at a lower velocity than coarse-grained soils, but this is not always the case, for the presence of cementing material in the soil may greatly increase its resistance to scour. The bed material of a canal tends to consolidate with use and develop increased resistance to erosion. Water which carries abrasive material is more effective in eroding cohesive or consolidated materials. Table 20.1 lists approximate maximum permissible velocities in channels of various materials. A more sophisticated approach to the scour problem involves comparing the boundary shear stress (tractive force per unit area) with the permissible unit tractive force. Through such an approach it is possible to achieve a balanced design, i. e., to determine the bottom width and side slope of the channel such that scour of bottom and sides is equally unlikely.

If the channel slope too gradual the velocity may be so low that growth of aquatic plants will reduce the hydraulic efficiency of the channel. Moreover, suspended sediment in the water may be deposited. Consequently, design velocities should be slightly less than the maximum permissible if topography permits.

Earth canals are generally trapezoidal, with side slopes determined by the stability of the bank material. The determination of stable slopes uses the procedures described for earth dams. Table 20.2 lists typical side slopes for unlined canals in various materials.

Table 20.1 Maximum permissible velocities in canals and flumes

| Channel material | English units | | SI metric units | |
|---|---|---|---|---|
| | Velocity (ft/sec) | | Velocity (m/sec) | |
| | Clear water | Water with abrasive sediment | Clear water | Water with abrasive sediment |
| Fine sand | 1.5 | 1.5 | 0.45 | 0.45 |
| Silt loam | 2.0 | 2.0 | 0.60 | 0.60 |
| Fine gravel | 2.5 | 3.5 | 0.75 | 1.00 |
| Stiff clay | 4.0 | 3.0 | 1.2 | 0.90 |
| Coarse gravel | 4.0 | 6.0 | 1.2 | 1.8 |
| Shale, hardpan | 6.0 | 5.0 | 1.8 | 1.5 |
| Steel | * | 8.0 | * | 2.4 |
| Timber | 20.0 | 10.0 | 6.0 | 3.0 |
| Concrete | 40.0 | 12.0 | 12.0 | 3.6 |

* Limited only by possible cavitation.

Table 20.2 Typical side slopes for unlined canals

| Bank material | Slopes (horizontal: vertical) |
| --- | --- |
| Cut in firm rock | 1/4:1 |
| Cut in fissured rock | 1/2:1 |
| Cut in firm soil | 1:1 |
| Cut in fill in gravelly loam | 1.5:1 |
| Cut or fill in sandy soil | 2.5:1 |

Freeboard must be provided above the design water level as a precaution against accumulation of sediment in the canal, reduction in hydraulic efficiency by plant growth, wave action, settlement of the banks, and flow in excess of design quantities during storms. Economy in the cost of excavation and earthwork is achieved primarily by balancing cut and fill. It may, however, be advantageous to borrow or waste where the haul distance is great. On side-hill locations the canal may be quite deep in order to balance cut and fill. Typical canal sections are shown in Fig.20.1.

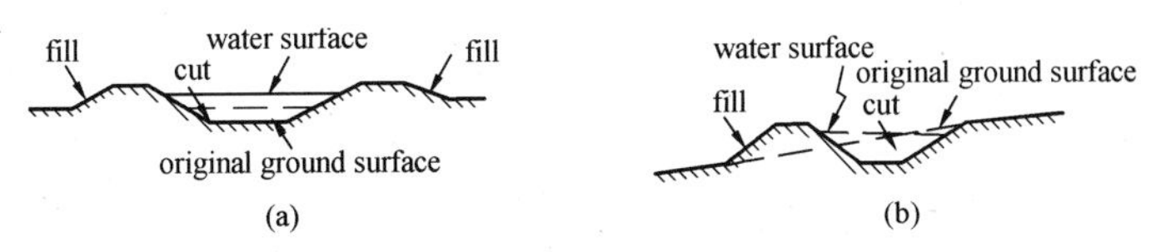


Fig. 20.1 Typical canal cross sections

(a) flat county; (b) side-hill location

If the water has high value and the soil in which the canal is constructed is quite permeable, it may be economical to provide a canal lining to reduce seepage from the canal. The rate of seepage from unlined canals is influenced chiefly by the character of the soil and the location of the groundwater level. Seepage rates may be measured by: ①ponding, ②inflow-outflow measurements, ③ seepage meter determinations. In the ponding method temporary watertight bulkheads are used to isolate a reach of canal. Water is admitted, and its rate of disappearance less an allowance for evaporation loss is the seepage rate. The inflow-outflow method is not reliable because a small error in one of these measurements will produce a relatively large error in the computed seepage rate. The seepage meter consists of an inverted metal cup with a face area of 2 $ft^2$ connected by tube to a flexible water bag that is initially filled with water. The cup is inserted face down into the canal bottom with the bag submerged in the canal water. A valve is opened permitting water to flow from the bag to the cup to satisfy seepage through the canal bottom. The rate of loss of water from the bag is indicative of the seepage rate. To determine an average seepage rate by this method, numerous tests at various points of the canal bottom are necessary.

Various types of linings are used to reduce seepage losses from canals. Clay, asphalt, cement mortar, and reinforced concrete have been used effectively. An effective and inexpensive lining is a buried membrane constructed by spraying asphalt over the sides and bottom of the channel and then placing a protective cover of about 6 in. (15 cm) of soil. The presence of fine sediment in the water may help to make the canal self-sealing. For important canals a concrete lining is usually most satisfactory because of its permanency. Reinforced concrete is used for canal linings in thicknesses of 2 to 8 in. (5 to 20 cm) depending on the size and importance of the canal. Standard reinforcement is 0.5 percent in the longitudinal direction and 0.2 percent in the transverse direction. Watertight construction joints are required at regular intervals. Mortar linings for small canals are often placed by guniting over steel mesh or by use of movable forms. For a concrete lining to be successful, the canal banks must be stable and well drained. Uplift under the lining may cause serious damage when the canal is empty. Seepage loss from properly lined canals may be as low as 0.05 ft/day (0.015 m/day). In addition to a reduction in seepage, lining may permit higher water velocities and smaller cross sections in the canal, with a resulting saving in cost.

## New Words and Expressions

1.downgrade *n.*, *a.* 下坡（度，的）
2.frequent *a.* 频繁的，时常发生的，经常的
3.route *n.* 路线；道路
4.tolerate *vt.* 允（容）许；承受
5.scour *n.* 冲刷，冲洗
6.cementing *n.* 粘结（牢）；胶结（合）
7.abrasive *a.* 磨蚀的，磨损的
8.cohesive *a.* 粘聚（性）的，（有）粘结（力）的
9.sophisticated *a.* 完美的，复杂的
10.tractive *a.* 牵引的；拖曳的
11.unlikely *a.* 未必的；不见得的
12.gradual *a.* 平缓的；不陡峭的
13.aquatic *a.* 水生的，水（上，中）的；*n.* 水生植物；水草
14.trapezoidal *a.* 不规则四方形的；梯形的
15.stable *a.* 稳定的；安定的
16.unlined *a.* 无衬砌的
17.earthwork *n.* 土方（工程）
18.haul n .搬运；运输
19.sidehill *n.* 山坡（边，侧）
20.temporary *a.* 暂时的；临时的
21.watertight *a.* 不漏（透）水，防渗的
22.bulkhead *n.* 隔板（墙，壁），围堰，护岸
23.isolate *vt.* 隔离，断开
24.reach *n.* 渠段，河段；可达到的距离
25.allowance *n.* 允许；限额；考虑
26.invert *vt.* 巅倒；（使）反向；反转
27.flexible *a.* 柔性的；可弯曲的，可活动的
28.indicative *a.* 指示的，表示特征的
29.asphalt *n.* 沥青，柏油
30.mortar *n.* 砂（灰，泥）浆
31.membrane *n.* 薄膜；隔板；止水墙
32.gravelly *a.* 砾质的
33.longitudinal *a.* 纵向的，轴向的
34.guniting *n.* 水泥枪；喷（射，灌）浆；喷射水泥砂浆
35.mesh *n.* 网格，网眼
36.tractive force （河流的）曳引力；推移力
37.hydraulic efficiency 输水能力；水力效率
38.balancing cut and fill 挖填平衡
39.canal lining 渠道衬砌；运河衬砌
40.face down 面朝下，口朝下（动词短语）
41.be indicative of 表现（示）出，有…征兆

## Reading Material Canals

Two types of conduits are used to convey water, the open channel and the pressure conduit. The open channel may take the form of a canal, flume, tunnel, or partly filled pipe, Open channels are characterized by a free water surface, in contrast to pressure conduits, which always flow full.

Fig. 20.2 below illustrates conditions for uniform flow of water in an open channel. Writing the energy equation between sections A and B gives

$$z_A + y_A + \frac{V_A^2}{2g} = z_B + y_B + \frac{V_B^2}{2g} + h_l \tag{20.1}$$

Where: $z$ is the elevation of the channel bottom above an arbitrary datum in feet; $y$ is the depth of flow; $V$ is the average velocity; $h_l$ is the head loss between A and B. Each term in the equation is in foot-pounds per pound (Newton meters per Newton) of water flowing, and hence, in feet (or meters). In uniform flow $y_A = y_B$ and $V_A = V_B$; hence $h_l = z_A - z_B = SL$, where $S$ is the slope of the energy grade line and $L$ is the distance between sections

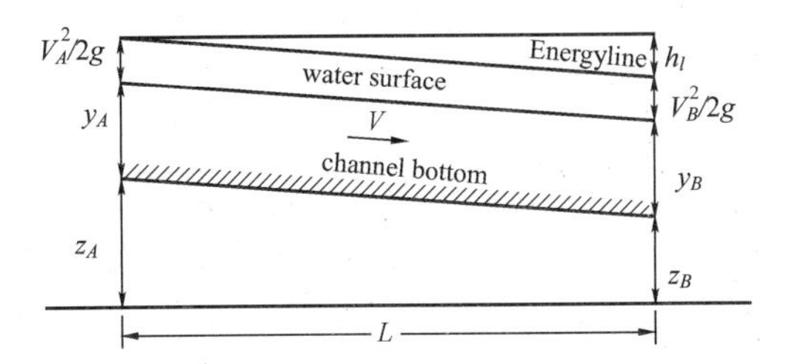


Fig. 20.2 Definition sketch for uniform flow in an open channel

Several equations are used to calculate the rate of flow in an open channel. The Chezy equation is

$$V = C\ (\ RS\ )^{1/2} \tag{20.2}$$

where: $V$ is the average velocity of flow; $C$ is a coefficient; $R$ is the hydraulic radius (cross-sectional area divided by wetted perimeter); $S$ is the slope of the energy grade line (equal to the slope of the water surface and also the channel bottom in uniform flow). The Chezy coefficient is most frequently expressed as

English units:

$$C = \frac{1.49}{n} R^{l/\theta} \tag{20.3}$$

SI metric units:

$$V = \frac{1.49}{n} R^{3/2} S^{1/2} \tag{20.4}$$

Combining Eqs. (20.2) and (20.3) results in what is commonly called the Manning equation:

English units:

$$C = \frac{1}{n} R^{L/B} \tag{20.5}$$

SI metric units:

$$V = \frac{1}{n} R^{1/3} S^{1/2} \tag{20.6}$$

which is applicable when the channel slope is less than about 0.10. Under such conditions the slant length of the channel does not differ materially from $L$. In Eq. 20.4 $V$ is expressed in feet (meters) per second and $R$ is expressed in feet (meters).

The factor $n$ in Eqs. (20.3) to (20.6) is the roughness coefficient of the channel (Table 20.3). If the roughness is not uniform across the channel width, and average value of $n$ must be selected or the channel may be treated as two or more contiguous channels, each having its own value of $n$. The discharge in each subdivision of the channel is computed independently, and the separate values are added to obtain the total flow. Natural channels with over-bank flood plains are often treated in this manner.

Various types of openchannel problems occur in engineering practice. Tables, nomo-graph, slide rules, and other devices have been prepared to simplify calculations and facilitate solution of these problems. The nomo-graph is applicable for SI units when $n$=0.013. It should be noted that $V$ and $Q$ are proportional to $1/n$ and $S$ proportional to $n^2$ so that values from the nomo-graph may be readily adjusted to any other value of $n$.

Table 20.3 Values of the roughness coefficient $n$

| Channel material | $n$ | Channel material | $n$ |
|---|---|---|---|
| Plastic, glass, draw tubing | 0.009 | Neat cement, smooth metal | 0.010 |
| Planed timber, asbestos pipe | 0.011 | Wrought iron, welded steel, canvas | 0.012 |
| Ordinary concrete, asphalted cast iron | 0.013 | Unplaned timber, vitrified clay | 0.014 |
| Cast-iron pipe | 0.015 | Riveted steel, brick | 0.016 |
| Rubble masonry | 0.017 | Smooth earth | 0.018 |
| Corrugated metal pipe | 0.022 | Firm gravel | 0.023 |
| Natural channels in good condition | 0.025 | Natural channels with stones and weeds | 0.035 |
| Very poor natural channels | 0.060 | | |

The problem of canal location is similar in many respects to highway location, but the solution may be more difficult since the slope of the canal bottom must be downgrade, and frequent changes in slope (and hence changes in section) should be avoided. Within the limitations of topography, the exact route of a canal is determined by the slopes that can be tolerated. Excessive slope may result in a velocity sufficient to cause erosion of the channel bottom or sides. The velocity at which scour will begin depends on the bed material and the shape of the channel cross section. Fine-grained soils generally scour at a lower velocity than coarse-grained soils, but this is not always the case, for the presence of cementing material in the soil may greatly increase its resistance to scour. The bed material of a canal tends to consolidate with use and develop increased resistance to erosion. Water which carries abrasive material is more effective in eroding cohesive or consolidated materials. Table 20.4 lists approximate maximum permissible velocities in channels of various materials.

If the channel slope is too gradual the velocity may be so low that growth of aquatic plants will reduce the hydraulic efficiency of the channel. Moreover, suspended sediment in the water may be deposited. Consequently, design velocities should be slightly less than the maximum permissible if topography permits.

Table 20.4 Maximum permissible velocities in canals and flumes

| Channel material | English units | | SI metric units | |
|---|---|---|---|---|
| | Velocity (ft/sec) | | Velocity (m/sec) | |
| | Clear Water | with water abrasive sediment | Clear Water | with water abrasive sediment |
| Fine sand | 1.5 | 1.5 | 0.45 | 0.45 |
| Silt loam | 2.0 | 2.0 | 0.6 | 0.6 |
| Fine gravel | 2.5 | 3.5 | 0.75 | 1.0 |
| Stiff clay | 4.0 | 5.0 | 1.2 | 0.9 |
| Coarse gravel | 4.0 | 6.0 | 1.2 | 1.8 |
| Concrete | 40.0 | 12.0 | 12.0 | 3.6 |

Earth canals are generally trapezoidal, with side slopes determined by the stability of the bank material. The determination of stable slopes uses the procedures described for earth dams. Table 20.5 lists typical side slopes for unlined canals in various materials.

Table 20.5 Typical side slopes for unlined canals

| Bank material | Slopes(horizontal: vertical) |
|---|---|
| Cut in firm rock | 1/4:1 |
| Cut in fissured rock | 1/2:1 |
| Cut in firm soil | 1:1 |
| Cut or fill in gravelly loam | 3/2:1 |
| Cut or fill in sandy soil | 5/2:1 |

Freeboard must be provided above the design water level as a precaution, a giant accumulation of sediment in the canal, reduction in hydraulic efficiency by plant growth, wave action, settlement of the banks, and flow in excess of design quantities during storms.

If the water has high value and the soil in which the canal is constructed is quite permeable, it may be economical to provide a canal lining to reduce seepage from the canal. The rate of seepage from unlined canals is influenced chiefly by the character of the soil and the location of the groundwater level.

## New Words and Expressions

1.arbitrary *a.*任意的，独立的

2.datum *n.* (*pl.* data) 数据，基准（点，线，面）

3.radius *n.* (*pl.* radii or -es) 半径，范围，辐射光线；hydraulic~水力半径

4.perimeter *n.* 周，周长

5.slant *vt.* 使倾斜；*vi.* (towards)倾向；*n.* 倾斜

6.contiguous *a.* 邻接的，接触的

7.subdivision *n.* 分部（段，支），小（节），部分

8.overbank *n.* 河滩

9.nomo-gram *n.* (or nomo-graph) 列线（线示，诺谟）图，图解

10.facilitate *vt.* 便于，促进，简化，减轻…困难

11.downgrade *n.* 下坡，退步；*ad.* 降下；*n.* 下坡路

12.abrasive *a.* 有研磨（磨损）作用的；*n.* 研磨剂

13.trapezoid *n.* （英）不规则四边形；（美）梯形；-al *a.*

14.precaution *n.*，*vt.* 预防，保护

15.energy grade line 能坡线

16. SI=Sysetime International d' Unites 国际单位制，公制

17.with use 由于经常使用

18.maximum permissible velocity in channel 渠道不冲流速

# Lesson 21 Land Drainage

Land drainage removes excess surface water from an area or lowers the groundwater below the root zone to improve plant growth or reduce the accumulation of soil salts. Land-drainage systems have many features in common with municipal storm drainage of surface water, at a considerable saving in cost over that of buried pipe. Open ditches, which are less objectionable in rural areas than in cities, are widely used under suitable soil conditions ditches may also serve to lower the water table. However, closely spaced open ditches will interfere with farm operations, and the more common method of draining excess soil water is by use of buried drains. Drains usually empty into ditches, although the modern tendency is to use large pipe in lieu of ditches where possible. This frees extra land for cultivation and does away with unsightly and sometimes dangerous open ditches. Since land drainage is normally a problem in very flat or leveed land, a disposal work provided with tide gates and pumping equipment is often necessary for the final removal of the collected water.

Land drainage speeds up the runoff of water and, hence, increases peak flow downstream of the drained area. The consequences of this increase should be considered in the planning of drainage systems. Wetlands are important biological areas. They serve migratory waterfowl and, in coastal areas, as nursery grounds for many important commercial species of aquatic life. The consequence of draining such lands requires careful evaluation.

Land drainage is not as demanding in terms of hydrologic design as other types of drainage. The purpose of land drainage is to remove a volume of water in a reasonable time. Where sub-drainage is installed to remove excess water from irrigated land for salinity control, the volume of leaching water to be applied in each irrigation is known; and the drains should be capable of removing this volume in the interval between twice irrigation.

Drainage installed for removal of excess water from rainfall is typically designed to remove a specified quantity of water in 24-hr. Commonly known as the drainage modulus or drainage coefficient, recommended values are about 1 percent of the mean annual rainfall. The drainage modulus as defined above will usually be in the range of one-quarter to one-half of the 1-yr 24-hr rainfall and may be taken as an estimate of the infiltration from the 1-yr 24-hr rain less the water required to bring the soil to field capacity. If the drainage modulus for irrigated lands exceeds the capacity defined by the estimated quantities of leaching water, the higher value should be used. Commonly, however, the minimum tile size (usually 4 in, or 10 cm) will be more than adequate to handle the design flow.

Land protected by levees is often subject to excess water as a result of seepage from the river through or under the levee. The design flow for a drainage system in this case should be based on estimated seepage rates at river stages having a return period of 1 or 2 yr. If the high stages are expected to be concurrent with the rainy season the drainage modulus should be added to the seepage.

A ditch-drainage system consists of laterals, submains, and main ditches. Ditches are usually unlined. Small ditches may be constructed with special ditching machines, while larger ditches are often excavated with a dragline. Some very large ditches are constructed with floating dredges. Unless the excavated material (spoil) is needed as a levee to provide additional flow area in the ditch, it should be placed at least 15 ft (4.5m) back from the edge of the ditch so that its weight does not contribute to the instability of the ditch bank. Spoil banks decrease the cultivated area and prevent inflow of water from the land adjacent to the ditch. If possible, this spoil should be spread in a thin layer, but where this cannot be done, openings should be left in the spoil bank wherever a natural drainage channel intersects the ditch and at least every 500 ft (150m) along the ditch.

The slope available for drainage ditches is small, and the cross sections should approach the most efficient section as closely as possible. A trapezoidal cross section is most common, with side slopes not steeper than 1 on 1.5. Slopes of 1:2 or 1:3 are required in sandy soils. Occasionally, where the drainage water must be pumped, ditches are deliberately made with an inefficient section to create as much storage as possible to minimize peak pumping loads. The slope, alignment, and spacing of ditches are determined mainly by local topography. The minimum practicable slope is about 0.00005 (3 in. per mile, or 5 in per kilometer). The ditches generally follow natural depressions, but where possible are run along property lines. Lateral ditches rarely need to be spaced more closely than 1/2 mi (800m) apart, although on level land a spacing of 1/4 mi (400m) may be necessary. With favorable land slopes a ditch spacing of 1 mi (1600m) may be feasible.

Ditches are usually between 6 and 12 ft (2 to 4m) deep. Where tile drainage is to be used, the lateral ditches must be deep enough to intercept the under drains which are to discharge into it. Similarly sub-main and main ditches must be deep enough to receive the flow of lesser ditches. If the terrace is flat and the ditches quite long, an excessive depth may be required for the main ditch and it may be advisable to divide the system into two or more portions to shorten ditch lengths. In muck or peat soil, considerable subsidence may occur when the water is drained from the soil, and ditches must be constructed proportionately deeper.

The basic procedure for the design of land-drainage systems is not much different from that for the design of municipal storm drains. The steps in design may be summarized as follows:

(1) Prepare a detailed contour map of the area. A contour interval of 1 ft (0.3m) is commonly necessary.

(2) Select the location of the system outlet. If several outlets are possible, an economy study of the alternatives may be necessary.

(3) Determine the drainage modulus for under-drains and estimate the amount of water the ditches will intercept.

(4) Lay out a system of ditches (or pipe mains) of adequate size to carry the expected flows.

(5) Determine the proper depth for tile drains and plan the tile-drain layout. Field drains of the customary minimum size (4 in, or 10cm) will usually have adequate capacity, and it will be necessary to calculate only required sizes for mains and sub-mains.

(6) The first trial layout of mains may require revision after the plan for under-drains is completed. The entire system should be planned for minimum cost by use of shortest possible routes for pipes and ditches.

(7) Estimate project costs and proceed with legal steps necessary to undertake the project.

## New Words and Expressions

1.municipal *a.* 市的，市政的；-ity 自治区（市）

2.objectionable *a.* 该反对的，不适合的，不好的

3.lieu *n.* 场所，仅用于词组 "in lieu of（代替）"中

4.unsightly *a.* 难看的，丑的

5.migratory *a.* 迁移的，流动的

6.fowl *n.* 鸟，禽

7.nursery *n.* 苗圃（床），繁殖（培鱼）场，托儿所；rice ~ 水稻秧田

8.concurrent *a.* 同时发生的，共同作用的(with)

9.lateral *a.* 横（向）的，侧向的，交线的；-s *n.* 支渠，支沟

10.submain *n.* 分干渠(沟)

11.excavate *v.* 挖掘(穿、通)，开挖

12.dragline *n.* 牵引索，索式挖掘机

13.dredge *n.* 挖泥机（船）

14.spoil *v.* (spoilt or spoiled) 损害；*n.* 脏物，弃土；~ bank 弃土堤

15.occasion *n.* 时机，原因，理由；*vt.* 引起；-al *a.* 偶尔的，临时的；-ally *ad.* 偶尔，间或

16.deliberate *v.* 考虑，讨论；*a.* 谨慎的，从容的，故意的 -ly *ad.*；–ation *n.*

17.depression *n.* 降低（落），沉降，凹（洼地）

18.terrace *n.* 阶地

19.advisable *a.* 适当的，贤明的

20.muck *n.* 腐殖土，淤泥，软土

21.peat *n.* 泥炭土

22.subside vi.下沉，沉陷（淀），减退（少）；-nce *n.* 剩余水头

23.neat *a.* 整洁（齐）的，平（光）滑的；-ness *n.*; -ly *ad.*

24.revision *n.* 校订，修正

25.undertake *vt.* 承担，接受，着手

26.in common (with) 公有的，共同的，相象的

27.storm-drain systems （城市）暴雨排水系统

28.open ditches 明沟排水（网，系统）

29.to be used as （被）用作

30.to empty into… 排水入…

31.leveed land 垸地，围垦地

32.disposal works 排水枢纽

33.tide gate 挡潮闸

34.to speed up 加快，赶快

35.drainage modulus *n.* (*pl.* moduli) 排水模数，排水率

36.1-yr 24-hr rainfall 设计年最大 24 小时暴雨

37.the most efficient section 最优水力断面

38.the shortest possible route 最短可能路径法

39.to proceed (from) 从…出发，出自；-(to) 着手，继续进行；-(with) 继续进行

40.mi = mile 英里

41.in terms of… 依据…

## Reading Material Flow of groundwater

The flow of groundwater to a drainpipe or ditch is governed by the same factors controlling the flow to a well. Both ditches and drains create a water table like that shown in Fig. 21.1. The cone of depression about a well becomes a trough along the line of the drain. The spacing of drains must be such that the water table at its highest point between drains does not interfere with plant growth. The water table should be below the root zone to permit aeration of the soil. A high water content also reduces soil temperature and retards plant growth. The necessary depth to the water table depends upon the crop, the soil type, the source of the water, and the salinity of the water. In humid regions, a depth of 2 to 3 ft (0.6 to 1.0m) is satisfactory for most crops except orchards, vineyards, and alfalfa, which require a depth of 3 to 5 ft (1 to 1.5m). In general, the water table should be lowered more in heavy clay soils than in light sandy soils. For land under irrigation the depth of the water table should be greater, 3 to 5 ft (1 to 1.5m) for ordinary crops and good-quality water, and 5 to 8 ft (1.5 to 2.5m) for deep-rooted crops. If the water contains excessive salts, the minimum depth for any crop should be about 4 ft (1.2m). It is not desirable to lower the water table far below the recommended minimum depth, for this deprives the plants of capillary moisture needed during the dry season. Where deep ditches are part of the drainage system, check dams should be provided to maintain the water level in the ditch during low flow periods in order to avoid excessive lowering of the water table near the ditch.

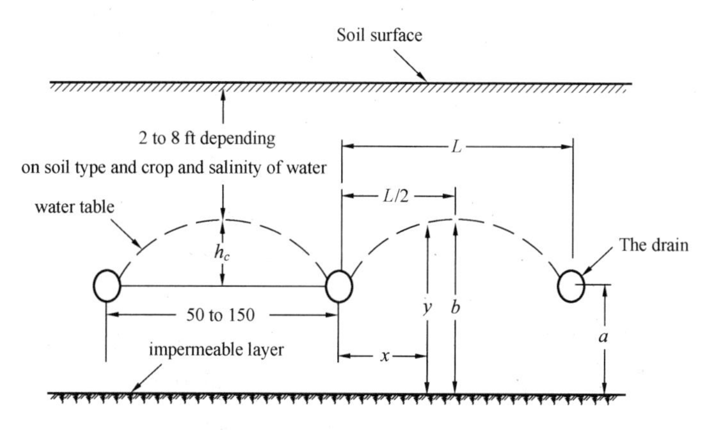


Fig. 21.1 Groundwater in a drained field

The variation in water-table elevation in the drained land should be not much over 1 ft (30cm). This means that drains should be placed about 1 ft (30cm) below the desired maximum groundwater level. The spacing of the drains must be such that, with the hydraulic gradient thus established (approximately 1 ft, 30cm in half the drain spacing), the drains will remove sufficient water from the soil. In homogeneous soil the spacing can be determined by an analysis similar to that for a well in a homogeneous aquifer. In Fig. 21.1 the hydraulic gradient at distance $x$ from the drain is $dy/dx$. Assuming the flow lines are parallel for a unit length of drain, the cross-sectional area of flow is $y$. Assuming that the flow $q$ toward the drain is inversely proportional to distance from the drain, i. e. , when $x = L/2$, $q = 0$ and when $x = 0$, $q = (1/2)\, q_0$, where $q_0$ is the design flow per foot of drain, the Darcy equation for the horizontal flow at x may be written.

$$q = \frac{2}{L}\left(\frac{L}{2} - x\right)\frac{q_0}{2} = Ky\frac{\mathrm{d}y}{\mathrm{d}x} \tag{21.1}$$

Where: $L$ is the drain spacing and $K$ is the coefficient of permeability. Transforming this equation and integrating.

$$\frac{q_0}{2KL}(L-2x)\mathrm{d}x = y\mathrm{d}y \tag{21.2}$$

When $x$=0, $y$ =a and hence $C$=$-a^2/2$. Substituting this value for $C$ in Eq.(21.3) and solving for $K$

$$\frac{q_0}{2KL}(Lx-\frac{2x^2}{2}) = \frac{y^2}{2}+C \tag{21.3}$$

$$K=\frac{q_0(Lx-x^2)}{L(y^2-a^2)} \tag{21.4}$$

When $x$=$L$/2, $y$ = $h$, hence

$$L=\frac{4K(b^2-a^2)}{q_0} \tag{21.5}$$

The assumption of an inverse linear variation in $q$ with distance from a drain is only an approximation. However, Aronovici and Donnan show good agreement between Eq.(21.5) and field tests in relatively permeable soils. The lower the soil permeability, the greater the curvature of the water table near the drain and the less accurate the assumption. Dagan and Hammad present more sophisticated methods for determining depth and spacing of drains which would be more appropriate for soils of low permeability. Hammad also discusses the effect of evaporation from the soil which may reduce the drainage requirement. If no restricting layer is present in the soil Sutton suggests that should be taken equal to the tile depth.

The most important single factor in determining drain spacing is the soil permeability. Since drainage normally involves only a shallow depth of soil, small-diameter piezometers or open auger holes may serve as test “wells”. It is convenient to calculate the permeability from the rate of rise of water in such a hole after a portion of the water is pumped out.

Ditches, because of their large dimensions are somewhat more effective than pipe drains in removing soil moisture. However, the increased effectiveness is not sufficient to permit ditch spacing which will not interfere with farm operations. Moreover, ditches reduce the area available for farming. Hence, open ditches for removal of soil water are used only infrequently, and then only in the most permeable soils.

## New Words and Expressions

1.cone *n.* 锥形物，弹头，漏斗；*vt.* 使成锥形

2.trough *n.* 槽，低谷

3.aeration *n.* 通气，松散

4.deprive *vt.* 剥夺

5.homogeneous *a.* 同类的，均匀的，齐次的

6.aquifer *n.* 含水层

7.integrate *v.* 求积分，总和

8.curvature *n.* 曲率

9.piezometer *n.* 流体压力计，测压管，地下水位计

10.auger *n.* 螺旋钻，土钻

11.dimension *n.* 尺寸，维数，大小

12.humid region 湿润地区

13.transform *v.*，*n.* 变换，重排

14.coefficient of permeability 渗透系数，导水率，水力传导度

# Lesson 22 Methods of Applications of Drip Irrigation Systems

Selection of the type and specifications of the drip irrigation system depends on several parameters-water quality, soil type, the specific crop and climatic conditions. In the case of sandy soil land brackish (saline) water, the drippers need to be located very close to the plant. The plant will then be able to establish a well developed root system in the zone where water and nutrients are available, while the concentration of salts will be minimal. However, the volume of the root system will be small, and therefore frequent irrigation will be required. The amount of water discharged in each cycle of irrigation (even when the field is irrigated more than once a day) must be sufficient to provide appropriate leaching of the active root system-at least 1.5~2 $L/h$ from each dripper.

Spacing between drippers is usually determined by the stand (density) of plants in the field. Only afterwards are the soil type and water quality taken in account. For example, in the Arava valley tomatoes are planted at intervals of 0.5 meter between vines along the bed, so that the spacing between drippers is also 0.5 m [Fig.22.1 (a)]. When peppers are irrigated with brackish water one dripper line is attached to each line of plants, and the lateral is placed within a furrow 5~10 cm deep. Two plants are planted next to each dripper. On each bed 2 or 3 laterals are installed, and the distance between the drippers is 40 to 50 cm[Fig.22.1(b)]. When water of good quality is applied it is possible to install a single dripper line between a pair of two lines of pepper. The average distance between the pepper plants and the dripper is about 25 cm. Cultivation of onions or carrots calls for wetting of the whole area of the bed. This condition is achieved by installing two dripper lines for each bed [Fig. 22.1(c)]. The dripper discharge rate is determined by the soil type $4L/h$ in sandy soil and $2L/h$ in loess.

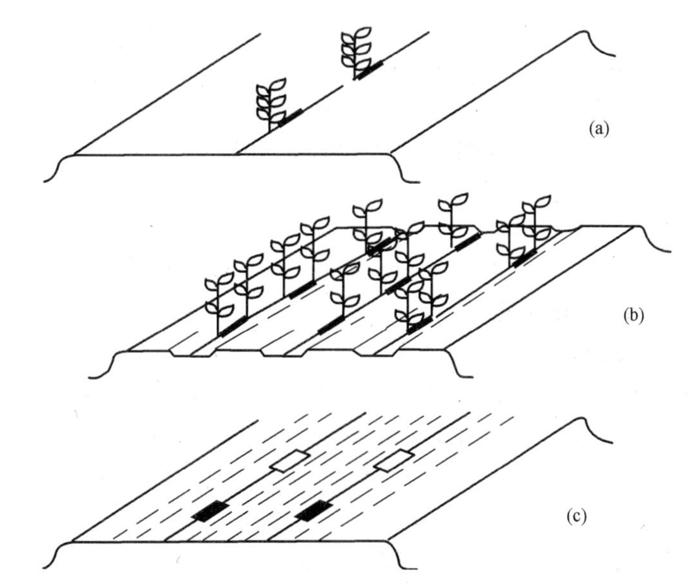


Fig. 22.1 Different methods of installation of irrigation systems in the Arava district

(a) tomatoes; (b) pepper; (c) onion

In regions where annual precipitation is over 150 mm drippers are installed at even intervals along the laterals regardless of the location of the tree, since the development of the root system is enhanced by the winter rains and it is necessary to supply water to the expanded root system.

Installation of long laterals is more economical since it allows the user to reduce the number of distributing units and simplifies the deployment of the system in the field. However, extending the length of the dripper lines requires the application of pressure compensated drippers which provide a uniform water supply along the line. Drip irrigation systems with long dripper lines are useful for crops cultivated over large areas, like cotton and maize (corn).

In orchards in arid zones the drippers are installed in close proximity to each tree. In order to supply palm trees with the appropriate amount of water in the Arava valley, about 10 drippers, each with a discharge rate of 8 *L/h* are installed next to each tree. The spacing between the drippers is about 30 cm along a line of 3 meters. In mango plantations 5 drippers of 8 *L/h* are installed along a dripper line segment of 1 m. Under these conditions the development of the root system is determined by the volume of the leached zone formed by the irrigation regime.

When using long laterals special attention should be given to the following parameters: the time required to fill the system relative to the length of the irrigation cycle, the flow rate of the water toward the end of the dripper lines, and drainage of the dripper lines.

### 1 Filing Time

In long dripper lines with tube diameter of 16 or 20 mm, a relatively large volume of water is needed to fill up the whole system and to generate the pressure required to regulate the discharge rate. When the irrigation system starts to fill up it often happens that water is discharged from the drippers in the beginning of the line, sometimes in relatively large quantities, even before the water reaches the drippers at the end of the line. When the intervals between irrigation cycles are long, i.e. more than three days, a filling time of abut 10 minutes is tolerable; however, when irrigation are applied at shorter intervals, say one or two days, the filling time can reach about 16% of the irrigation cycle. In order to reduce the time required for filling the system, valves and filters with large diameters should be installed to ensure a large volume of flow.

### 2 Flow Velocities

The flow rate in dripper lines in the range of 400 to 600 meters tends to drop and there is therefore a higher risk of clogging of the drippers by particles carried by the water. To overcome this problem a special device, the lateral flusher has been developed. The flushing device is installed at the end of the line, and makes it possible to wash the lines by applying a rapid stream of water. Flushing of the pipes continues until a steady water pressure is built up in the system, after which the device is shut off.

### 3 Dripper Lines Drainage

If distributers are installed in the lowest side of a slopping field, a problem of excess wetting of the lower section of a field may develop at the end of each irrigation cycle. Therefore, in sloping terrain the distributers must be installed on the high ground.

## New Words and Expressions

1.specifications n.（尺寸）规格，技术要求，规范工程
2.brackish a.（略）有盐味的，稍（微）碱的，碱化的，（轻度）盐渍的；~water 微（半）碱水
3.discharge v. 卸下，放（排，流）出，排放
4.be sufficient for 足以满足
5.be sufficient to + inf. 足够（以）做
6.stand n. 位置，站立，台灯
7.only conj. 但是，可是，不过；a. 惟一的，仅有的，最好的；ad. 仅仅，只
8.vine n. 藤，蔓
9.pepper n. 胡（辣，花）椒
10.onion n. 洋葱，葱头
11.carrot n. 胡萝卜，屑，政治欺骗
12.loess n. 黄土，大孔性土
13.in close proximity to 在极接近于…之处，紧靠着，非常接近于
14.palm n. 棕榈（叶），手心
15.plantation n. 种（栽）植，人造林，大农场，庄园；(pl.)新开地，创设，灌输
16.segment n. 部分，线段，扇形
17.regardless a. 不注意的，不关心的；ad.无论如何；~of 不管，不关心，无论，与…无关
18.enhance vt. 增加（强），提高，增长（高，加）
19.expand v. 扩展，延伸，伸长，发展
20.rain n., vi. 雨（水，天），下雨；(pl.)阵雨，季节雨，雨季，电子流
21.simplify vt. 简化，化（精）简，单一（纯）化，使单纯，使易懂（做）
22.deployment n. 配置，部署，调度，使用，采用，利用，推广应用
23.tolerable a. 允许的，可容忍的；相当好的，过得去的
24.mango n. 芒果，芒果树

## Reading Material Sprinkler System Design Requirements

The important factors in the success of sprinkler irrigation systems are first, the correct design, and second, the efficient operation of the designed system. The basic information necessary for the design of a farm irrigation system is obtained from four sources, namely, the soil, the water supply, the crop to be irrigated, and the climate.

Information concerning soils includes the soil type, depth, texture, permeability, and available water-holding capacity of the root zone. Necessary water supply concerns the location of the water delivery point in relation to the fields to be irrigated, the quantity of water available, and the delivery schedule. The maximum consumptive use of water per day, the root-zone depth, and the peculiarities of irrigation necessary to be taken into account in the irrigation system are obtained from a knowledge of the requirements of the crop to be grown. Climatological information includes the natural precipitation and wind velocities and direction. All this information must be compiled in one form or another before starting to design a sprinkler system.

The performance requirements of a sprinkler system include applying water at a rate that will not cause runoff from the area irrigated. Water should be applied at such a rate that high water-application efficiency is obtained and in such a manner that high water-distribution efficiency results.

The sprinkler system must have the capacity to meet the peak water-use demands of each crop during the irrigation season. Allowance in capacity must be made for unavoidable water losses by evaporation, interception, and some deep percolation.

When a system is designed for supplemental irrigation or protective purposes, it should have a capacity to apply the necessary depth of water to the design area in a specified time. The cost of the system should be consistent with the insurance values involved.

There should not be more than 20 percent variation in the depth of water applied to any part of the design area. To obtain reasonably uniform distribution the pressure in the lateral lines should not vary more than 20 percent. Variations in pressure occur as the result of friction loss in pipes and elevation changes in main lines or lateral lines. It may be necessary to control pressures with valves.

A sprinkler system must apply water so that it will not cause physical damage to the crop. In orchards, high-velocity streams of water from sprinkler nozzles have bruised growing apples and peaches when sprinklers have been placed too close to the trees. Also, in crops having fine seedling plants, a fine spray must be applied, or the plants will be beaten into the ground. Such a spray requires high pressures to break up the water drops at the nozzle.

Droplets must be small enough so that the soil is not damaged. Some soils will puddle under the impact of large droplets, causing the soil to crust. Smaller nozzles operated at higher pressures reduce drop size. Acceptable droplet size can be obtained using the following combination of nozzle sizes and pressure heads.

(3.175 to 4.75mm)——25 to 35m

(4.75 to 6.35mm)——30 to 40m

(6.35 to 9.5mm)——35 to 50m

Two nozzle sprinklers will produce acceptable breakup of drops at 3.5 meters less pressure head than single nozzle sprinklers.

Wind can be critical, interfering excessively with distribution of the water. It is difficult to obtain good distribution of water in winds in excess of 15 kilometers per hour. Lateral lines should be set at an angle to the prevailing wind. It may be necessary to decrease the spacing between sprinklers and between lateral lines as much as 40 percent to obtain satisfactory distribution in the windy condition.

When used in practical field spacing with selected operating pressures, the sprinkler chosen must give satisfactory moisture distribution.

If a pump is necessary it must be picked on the basis of the maximum operating conditions of pressure head and flow and must not overload at minimum operating conditions.

Careful attention should be given to the planning of the sprinkler system. To select the best combination of power plant, sprinkler system, and operating schedule requires skill and experience.

## New Words and Expressions

1.peculiarity *n.* 特质（征，性，色），奇特（异）

2.climatological *a.* 气候的，气象的

3.interception *n.* 拦截，相交，窃听

4.bruise *v., n.*撞（跌，打，碰，擦）伤，损伤，捣碎

5.peach *n.* 桃，桃红色

# Lesson 23 Hydraulic Turbines

## 1 Introduction

Power may be developed from water by three fundamental processes: by action of its weight, of its pressure, or of its velocity, or by a combination of any or all three. In modern practice the Pelton or impulse wheel is the only type which obtains power by a single process the action of one or more high-velocity jets. This type of wheel is usually found in high-head developments.

There has been practically no increase in the efficiency of hydraulic turbines since about 1925, when maximum efficiencies reached 93% or more. As far as maximum efficiency is concerned, the hydraulic turbine has about reached the practicable limit of development. Nevertheless, in recent years, there has been a rapid and marked increase in the physical size and horsepower capacity of individual units.

In addition, there has been considerable research into the cause and prevention of cavitation, which allows the advantages of higher specific speeds to be obtained at higher heads than formerly were considered advisable. The net effect of this progress with larger units, higher specific speed, and simplification and improvements in design has been to retain for the hydraulic turbine the important place which it has long held at one of the most important prime movers.

## 2 Types of Hydraulic Turbines

Hydraulic turbines may be grouped in two general classes: the impulse type which utilizes the kinetic energy of a high-velocity jet which acts upon only a small part of the circumference at any instant, and the reaction type which develops power from the combined action of pressure and velocity of the water that completely fills the runner and water passages. The reaction group is divided into two general types: the Francis, sometimes called the reaction type, and the propeller type. The propeller class is also further subdivided into the fixed-blade propeller type, and the adjustable-blade type of which the Kaplan is representative.

### 2.1 Impulse Wheels

With the impulse wheel the potential energy of the water in the penstock is transformed into kinetic energy in a jet issuing from the orifice of a nozzle. This jet discharge freely into the atmosphere inside the wheel housing and strikes against the bowl-shaped buckets of the runner. At each revolution the bucket enters, passes through, and passes out of the jet, during which time it receives the full impact force of the jet. This produces a rapid hammer blow upon the bucket. At the same time the bucket is subjected to the centrifugal force tending to separate the bucket from its disk. On account of the stresses so produced and also the scouring effect of the water flowing over the working surface of the bowl, material of high quality of resistance against hydraulic wear and fatigue is required. Only for very low heads can cast iron be employed. Bronze and annealed cast steel are normally used.

### 2.2 Francis Runners

With the Francis type the water enters from a casing or flume with a relatively low velocity, passes through guide vanes or gates located around the circumference, and flows through the runner, from which it discharges into a draft tube sealed below the tail-water level. All the water passages are completely filled with water, which acts upon the whole circumference of the runner. Only a portion of the power is derived from the dynamic action due to the velocity of the water, a large part of the power being obtained from the difference in pressure acting on the front and back of the runner buckets. The draft rube allows maximum utilization of the available head, both because of the suction created below the runner by the vertical column of water and because the outlet of the draft tube is larger than the throat just below the runner, thus utilizing a part of the kinetic energy of the water leaving the runner blades.

### 2.3 Propeller Runners

Inherently suitable for low-head developments, the propeller-type unit has effected marked economics within the range of head to which it is adapted. The higher speed of this type of turbine results in a lower-cost generator and somewhat smaller powerhouse substructure and superstructure. Propeller-type runners for low heads and small outputs are sometimes constructed of cast iron. For heads above 20 ft, they are made of cast steel, a much more reliable material. Large-diameter propellers may have individual blades fastened to the hub.

### 2.4 Adjustable-blade Runners

The adjustable-blade propeller type is a development from the fixed-blade propeller wheel. One of the best –known units of this type is the Kaplan unit, in which the blades may be rotated to the most efficient angle by a hydraulic servomotor. A cam on the governor is used to cause the blade angle to change with the gate position so that high efficiency is always obtained at almost any percentage of full load.

By reason of its high efficiency at all gate openings, the adjustable-blade propeller-type unit is particularly applicable to low-head developments where conditions are such that the units must be operated at varying load and varying head. Capital cost and maintenance for such units are necessarily higher than for fixed-blade propeller-type units operated at the point of maximum efficiency.

## New Words and Expressions

1.turbine *n.* 叶（汽，水）轮机；
hydrulic turbine 水轮机
2.Pelton wheel 培尔顿式水轮
3.impulse 冲击，冲动；
impulse wheel 冲击式水轮
4.high-velocity jet 高速射流
5.high-head development 高水头电站
6.horsepower *n.* 马力
7.units *n.* 机组
8.cavitation *n.* 空蚀
9.specific speed 比转速
10.simplification 简化
11.prime *a.* 原始的，最初的；
prime mover 原动机

12.kinetic *a.* 运动的，能动的；
kinetic energy 动能
13.circumference *n.* 轮周，周围
14.instant *n.* 时刻，瞬时
15.reaction type 反击式
16.runner 转轮
17.water passage 过水道
18.Francis type 弗朗西斯式
19.propeller type 旋浆式
20.blade *n.* 叶片
21.fixed-blade 定轮叶的，叶片固定的
22.adjustable-blade 转叶的，叶片可调的
23.Kaplan type 卡普兰式
24.bucket *n.* 戽斗，轮叶
25.impact *vt.* 冲击
26.centrifugal force 离心力
27.disk *n.* 转轮，圆盘
28.hydraulic wear 水力磨损
29.fatigue *n.* 疲劳
30.cast iron 铸铁
31.bronze *n.* 青铜
32.cast steel 铸钢
33.anneal *v.*，*n.* 韧化（退火），热处理
34.casing (or spiral case) 蜗壳
35.flume *n.* 水槽
36.guide vanes 导叶
37.draft tube 尾水管
38.seal *n.* 止水，密封
39.dynamic action 动力作用
40.runner bucket 转轮轮叶
41.available head 可（利）用（的）水头
42.suction *n.* 吸力，负压
43.throat *n.* 喉管
44.low-head development *n.* 低水头电站
45.powerhouse *n.* 发电厂房
46.substructure *n.* 水（地）下结构
47.superstructure *n.* 水（地）上结构
48.hub *n.* 轮毂
49.fasten *v.* 固定，拴紧
50.servomotor *n.* 接力器，伺服器
51.cam *n.* 凸轮
52.capital cost 造价，投资费用
53.maintenance *n.* 维护
54.babbit *n.* 巴氏合金、乌金
55.steady *a.* 稳定的
56.bearing *n.* 轴承
57.stuffing *n.* 填入，填塞
58.shift *n.*，*v.* 变速，滑移
59.actuating *n.* 启动，操纵
60.pit *n.* 坑
61.liner *n.* 衬砌
62.vane *n.* 叶片，叶轮；
guide vane 导叶
63.stem *n.* 柄，杆

# Reading Material Turbines

Hydraulic turbines are the machines which convert hydraulic energy into mechanical energy. The mechanical energy developed by a turbine is used in running an electric generator, which is directly coupled to the shaft of the turbine. The generator thus develops electric power which is also sometimes known as hydroelectric power. In general, a water turbine consists of a wheel called runner which is provided with specially designed blades or buckets. The water possessing large hydraulic energy strikes the runner and causes it to rotate.

Hydraulic turbines may be classified under two heads: ①Impulse or velocity turbines; ②Reaction or pressure turbines.

## 1 Impulse Turbine

In the case of impulse turbine, all the available potential energy or head is converted into kinetic energy or velocity head by passing it through a contracting nozzle or by guide vanes before it strikes the buckets of turbine. The wheel revolves free in air and water is in contact with only a part of wheel at a time. The pressure of water all along is atmospheric. In order to prevent splashing and to guide the water discharged from the buckets to the tailrace, a casing is provided. Thus, an impulse turbine has the following characteristic features:

(1) The wheel passage is not completely filled.

(2) The water acting on the wheel vanes is under atmospheric pressure.

(3) The water is supplied at a few points at the periphery of the wheel (usually one point, but occasionally two or more points).

(4) Energy applied to the wheel is wholly kinetic.

An impulse turbine is essentially a low-speed wheel and is used for relatively high heads, ordinarily beginning in the neighborhood of 150 m, and so far being limited to about 1000m. Though a head of over 1770m has been utilized by impulse wheels in Switzerland. Examples of the impulse turbines are: Pelton wheel, Turgo-impulse wheel, Girard turbine, Banki turbine, Jonval turbine etc. Out of these, the Pelton wheel is predominantly used at present.

## 2 Reaction Turbine

In the case of a reaction turbine, only a part of the available potential energy is converted into velocity head, at the entrance to the runner, and the balance that forms a substantial portion remains as a pressure head. The pressure at the inlet to the turbine is much higher than the pressure at the outlet, and it varies throughout the passage of water through the turbine. Major part of the power is derived from the difference in pressure acting on front and back of runner blades and only a minor part from the dynamic action of velocity. The water being under pressure, the entire flow from head race to tail race takes place in a closed conduit system and the atmospheric air has no access at any point. To summarize, a reaction turbine has the following characteristic features:

(1) The wheel passages are completely filled with water.

(2) The water acting on the wheel vanes is under pressure greater than atmospheric.

(3) The water enters all around the periphery of the wheel.

(4) Energy in the form of both pressure and kinetic is utilized by the wheel.

The examples for reaction turbines are: Fourneyron turbine, Tompson turbine, Propeller turbine, Francis turbine, Kaplan turbine, etc. Out of these, Francis and Kaplan turbines are predominantly used at present.

The turbines may also be classified according to the direction of flow of water in the runner, as follow: ①tangential flow turbine; ②radial flow turbine; ③axial flow turbine; ④mixed flow turbine.

In a tangential flow, the water strikes the runner along the tangent to the path of rotation of the runner, such as in the Pelton wheel. In a radial flow, the water flows along the radial direction and remains in a plane normal to the axis of rotation, as it passes through the runner. The radial flow may be inward (i. e. water entering the outer circumference and flowing radially inward) such as in the Francis turbine, Thomson turbine, or Girard turbine, or the flow may be outward (i. e. water entering at the center and flowing radially outward), such as in Fourneyron turbine or Boyden turbine. In an axial flow turbine or parallel flow turbine, water enters and leaves the runner along the direction parallel to the axis of the turbine shaft. Examples of this type of flow are in Jonval turbine, Girard turbine, propeller turbine and Kaplan turbine.

In the mixed flow turbine, water enters the runner at the outer periphery in the radial direction and leaves it at the center in the direction parallel to the axis of the shaft. Modern Francis turbine is one such example.

## New Words and Expressions

1.couple *vt.* 连接
2.shaft *n.* 轴；柄；柱身
3.kinetic *a.* 动力学的，运动的
4.vane *n.* 叶片；轮叶
5.race *n.* 水道；急流
6.casing *n.* 套管；机壳
7.periphery *n.* 周围；周边；圆周
8.neighborhood *n.* 附近（地区）；周围
9.predominantly *ad.* 主要地；突出地
10.tangential *a.* 切向的；切线的
11.tangent *n.* 切线；正切
12.impulse or velocity turbines 冲击式流速式水轮机
13.tail race 尾水渠
14.in the neighborhood of… 大约…；在…的附近
15.head race 引水渠
16.colsed conduit system 密闭管道系统
17.propeller turbine 轴流定桨式水轮机；螺桨式水轮机
18.tangential flow turbine 切流式水轮机
19.radial flow turbine 辐流（径流）式水轮机
20.axial flow turbine 轴流式水轮机
21.mixed flow turbine 混流式水轮机
22.parallel flow turbine 贯流式水轮机
23.Turgo-impulse wheel 土尔戈冲击式水轮机
24.Girard turbine 吉拉尔特式水轮机(早期的冲击式水轮机）
25.Banki turbine 班基式水轮机
26.Jonval turbine 琼维尔式水轮机
27.Fourneyron turbine 福内朗式水轮机
28.Francis turbine 弗朗西斯式水轮机
29.Tompson turbine 汤普森式水轮机
30.Kaplan turbine 卡普兰式水轮机
31.Thomson turbine 汤姆森式水轮机
32.Boyden turbine 博伊顿式水轮机

# Lesson 24 Hydro-electric Power

Faraday had shown that when a coil is rotated in a magnetic field, electricity is generated. Thus, in order to produce electrical energy, it is necessary that we should produce mechanical energy, which can be used to rotate the 'coil'. The mechanical energy is produced by running a prime mover (known as turbine) by the energy of fuels or flowing water. This mechanical power is converted into electrical power by electric generator which is directly coupled to the shaft of turbine and is thus run by turbine. The electrical power, which is consequently obtained at the terminals of the generator, is then transmitted to the area where it is to be used for doing work.

The plant or machinery which is required to produce electricity (i. e. Prime mover +Electric generator) is collectively known as power plant. The building, in which the entire machinery along with other auxiliary units is installed, is known as power house.

## 1 Thermal and Hydropower

As stated earlier, the turbine blades can be made to run by the energy of fuels or flowing water. When fuel is used to produce steam for running the steam turbine, then the power generated is known as Thermal power. The fuel which is to be used for generating steam may either be an ordinary fuel such as coal, fuel oil, etc., or atomic fuel or nuclear fuel. Coal is simply burnt to produce steam from water and is the simplest and oldest type of fuel. Diesel oil, etc. may also be used as fuels for producing steam. Atomic fuels such as uranium or thorium may also be used to produce steam. When conventional type of fuels such as coal, oil, etc. (called fossils) is used to produce steam for running the turbines, the power house is generally called an Ordinary thermal power station or Thermal power station. But when atomic fuel is used to produce steam, the power station, which is essentially a thermal power station, is called an Atomic power station or Nuclear power station. In an ordinary thermal power station, steam is produced in a water boiler, while in the atomic power station; the boiler is replaced by a nuclear reactor and steam generator for raising steam. The electric power generated in both these cases is known as Thermal power and the scheme is called Thermal power scheme.

But, when the energy of the flowing water is used to run the turbines, then the electricity generated is called hydroelectric power. This scheme is known as hydro scheme, and the power house is known as hydel power station or hydroelectric power station. In a hydro scheme, a certain quantity of water at a certain potential head is essentially made to flow through the turbines. The head causing flow runs the turbine blades, and thus producing electricity from the generator coupled to the turbine. In this chapter, we are concerned with hydel schemes only.

## 2 Classification of Hydel Plants

**Hydro-plants may be classified on the basis of hydraulic characteristics as follows:** ①Run-off river plants; ②Storage plants; ③Pumped storage plants; ④Tidal plants. They are described below:

**(1) Run-off River Plants.** These plants are those which utilize the minimum flow in a river having no appreciable pondage on its upstream side. A weir or a barrage is sometimes constructed across a river simply to raise and maintain the water level at a pre-determined level within narrow limits of fluctuations, either solely for the power plant or for some other purpose where the power plant may be incidental. Such a scheme is essentially a low head scheme and may be suitable only on a perennial river having sufficient dry weather flow of such a magnitude as to make the development worthwhile.

Run-off river plants generally have a very limited storage capacity, and can use water only when it comes. This small storage capacity is provided for meeting the hourly fluctuations of load. When the available discharge at site is more than the demand (during off-peak hours) the excess water is temporarily stored in the pond on the upstream side of the barrage, which is then utilized during the peak hours.

The various examples of run-off the river plant are: Ganguwal and Kotla power houses located on Nangal Hydel Channel, Mohammad Pur and Pathri power houses on Ganga Canal and Sarda power house on Sarda Canal.

The power stations constructed on irrigation channels at the sites of falls, also fall under this category of plants.

**(2) Storage Plants.** A storage plant is essentially having an upstream storage reservoir of sufficient size so as to permit, sufficient carryover storage from the monsoon season to the dry summer season, and thus to develop a firm flow substantially more than the minimum natural flow. In this scheme, a dam is constructed across the river and the power house may be located at the foot of the dam such as in Bhakra, Hirakud, Rihand projects etc. The power house may sometimes be located much away from the dam (on the downstream side). In such a case, the power house is located at the end of tunnels which carry water from the reservoir. The tunnels are connected to the power house machines by means of pressure pen-stocks which may either be underground (as in Mainthon and Koyna projects) or may be kept exposed (as in Kundah project).

When the power house is located near the dam, as is generally done in the low head installations; it is known as Concentrated fall hydroelectric development. But when the water is carried to the power house at a considerable distance from the dam through a canal, tunnel, or pen-stock; it is known as a Divided fall development.

**(3) Pumped Storage Plants.** A pumped storage plant generates power during peak hours, but during the off-peak hours, water is pumped back from the tail water pool to the headwater pool for future use. The pumps are run by some secondary power from some other plant in the system. The plant is thus primarily meant for assisting an existing thermal plant or some other hydel plant.

During peak hours, the water flows from the reservoir to the turbine and electricity is generated. During off-peak hours, the excess power is available from some other plant, and is utilized for pumping water from the tail pool to the head pool, This minor plant thus supplements the power of another major plant. In such a scheme, the same water is utilized again and again and no water is wasted.

For heads varying between 15m to 90m, reversible pump turbines have been devised, which can function both as a turbine as well as a pump. Such reversible turbines can work at relatively high efficiencies and can help in reducing the cost of such a plant. Similarly, the same electrical machine can be used both as a generator as well as a motor by reversing the poles. The provision of such a scheme helps considerably in improving the load factor of the power system.

**(4) Tidal Plants.** Tidal plants for generation of electric power are the recent and modern advancements, and essentially work on the principle that there is a rise in seawater during high tide period and a fall during the low ebb period. The water rises and falls twice a day; each fall cycle occupying about 12 hours and 25 minutes. The advantage of this rise and fall of water is taken in a tidal plant. In other words, the tidal range, i. e. the difference between high and low tide levels is utilized to generate power. This is accomplished by constructing a basin separated from the ocean by a partition wall and installing turbines in opening through this wall.

Water passes from the ocean to the basin during high tides, and thus running the turbines and generating electric power. During low tide, the water from the basin runs back to ocean, which can also be utilized to generate electric power, provided special turbines which can generate power for either direction of flow are installed. Such plants are useful at places where tidal range is high. Rance power station in France is an example of this type of power station. The tidal range at this place is of the order of 11 meters. This power house contains 9 units of 38,000 kW.

**Hydro-plants or Hydroelectric schemes may be classified on the basis of operating head on turbines as follows:** ①Low head scheme (head<15m); ②Medium head scheme (head varies between 15m to 60m); ③High head scheme (head>60m). They are described below:

**(1) Low Head Scheme.** A low head scheme is one which uses water head of less than 15 meters or so. A run off river plant is essentially a low head scheme. In this scheme, a weir or a barrage is constructed to raise the water level, and the power house is constructed either in continuation with the barrage or at some distance downstream of the barrage, where water is taken to the power house through an intake canal.

**(2) Medium Head Scheme.** A medium head scheme is one which used water head varying between 15 to 60 meters or so. This scheme is thus essentially a dam reservoir scheme, although the dam height is mediocre. This scheme is having features somewhere between low head scheme and high head scheme.

**(3) High Head Scheme.** A high head scheme is one which uses water head of more than 60m or so. A dam of sufficient height is, therefore, required to be constructed, so as to store water on the upstream side and to utilize this water throughout the year. High head schemes up to heights of 1,800 meters have been developed. The common examples of such a scheme are: Bhakra dam in (Punjab), Rihand dam in (U. P.), Hoover dam in (U. S. A.), etc.

The naturally available high falls can also be developed for generating electric power. The common examples of such power developments are: Jog Falls in India, and Niagara Falls in U. S. A.

# New Words and Expressions

1.coil *n.* 线圈，绕级；感应圈
2.collectively *ad.* 总起来说；共同地
3.auxiliary *a.* 辅助的；备用的
4.uranium *n.* 铀
5.thorium *n.* 钍
6.conventional *a.* 惯用的；传统的
7.fossil *n.* 矿物；化石
8.scheme *n.* 系统；设计图；方案
9.pondage *n.* 蓄水；调节容量
10.barrage *n.* 拦河坝，挡水建筑物
11.predetermine *vt.* 预定；对…先规定方向
12.fluctuation *n.* 波动；涨落
13.perennial *a.* 一年到头的；常年的
14.worth *n.* 性能；效用
15.hourly *a.*, *ad.* 每小时（地）；以钟点计算的（地）
16.off-peak *a.* 非高峰的；正常的
17.category *n.* 种类；类别；部门
18.carry-over *n.* 夹带；转移；滚进
19.substantially *ad.* 本质的，事实上
20.tunnel *n.* 隧道；隧洞；烟道
21.penstock *n.* 闸门；压力钢管；进水管
22.secondary *a.* 备用的；第二级的；再生的
23.supplement *vt.* 补充；增加；添加
24.partition *n.*, *vt.* 划分；分割
25.mediocre *a.* 普通的，中等的；无价值的
26.prime mover 原动机
27.run-off river plant 径流式电站
28.storage plant 蓄水式电站
29.pumped storage plant 抽水蓄能电站
30.load factor 负载系数
31.available discharge 可用流量；有效流量
32.off-peak hours 非峰荷时
33.fall under 列入…，属于…
34.carry-over storage 年调节库容
35.intake canal 进水渠
36.Faraday 法拉第
37.Ganguwal and Kotla power houses 冈古瓦尔和科拉水电站
38.Nangal Hydel channel 楠加尔·海得尔运河
39.Mohammad Pur and Pathri Power houses on Ganga Canal 恒河的穆罕默德·普尔和帕特里水电站
40.Sarda power house on Sarda 萨尔达运河的萨尔达水电站
41.Bhakra, Hirakud, Rihand projects 巴克拉，希陶库德，里亨德工程
42.Maithon and Koyna projects 迈吞和高勒工程
43.Kundah project 孔达工程
44.Rance power station in France 法国朗斯电站
45.Bhakra dam in Punjab 旁遮普省的巴克拉大坝（印度）
46.Rihand dam in U. P. 印度北方邦的里亨德大坝
47.Jog Falls in India 印度的乔客瀑布
48.Niagara Falls in U. S. A. 美国尼亚加拉瀑布

# Reading Material China's Hydropower Potential

## 1 Five surveys in four decades

There were five surveys conducted in the past four decades about the hydropower potential in China on a national basis.

The first survey was carried out during the period 1943~1944, during the World War Ⅱ. It was aimed at the evaluation of theoretical hydropower potential in China for the first time. The results given were 232,000 MW as calculated on average river discharge ($Q_{av}$), and 74,600 MW on the river discharge at 95% of the time ($Q_{95}$). The results were inaccurate, because the basic data used were inaccurate as well as incomplete.

The second survey was carried out during the period 1946~1947, and was supplemented in 1949. The final figure given was 149,000 MW. This figure more or less represented the exploitable hydropower potential as esteemed at that time. However, the basic data were still inaccurate and incomplete.

After the founding of the People's Republic , a third survey on hydropower potential was carried out in the period 1954~1955. This time, again, it was aimed at the evaluation of theoretical potential. The third survey greatly outmatched the previous two. It is not only because of the accumulation of basic data, but also because of the scope of survey deliberately extended to cover as many rivers as possible. There were 1998 rivers included in the survey, covering about 70% of the national territory of China. The total river length investigated amounted to 226,000 kilometers. The average annual runoff of these rivers amounted to 2,600 cubic kilometers. The result thus obtained was 540,000 MW calculated on average river discharge, and it was estimated that the exploitable hydropower potential in China would be around 300,000 MW.

The fourth survey was carried out in 1958. It was actually a general review of the work accomplished from 1951 to 1955, with adjustments and corrections here and there. The result was 580,000 MW.

The fifth survey was carried out in the period 1977~1978. A large number of personnel was assigned to the survey, and a very exhaustive work was accomplished. The following points are worth mentioning for the fifth survey:

(1) Hydrological records were available at much longer time periods, and many more hydrological stations have been established since 1949.

(2) More reliable and accurate topographic maps based on aerial surveying were available for nearly all parts of the country.

(3) A large number of river basin planning reports were available as a result of intensive studies done to nearly all the major river systems in China since 1949.

(4) The survey was aimed at both the theoretical hydropower potential and the exploitable hydropower potential, with more stress put on the latter.

The firth survey gave the following results:

(1) Total theoretical hydropower potential in China: 676,000 MW, corresponding to an annual generation of $5{,}922\times10^{12}$W·h.

(2) Total exploitable hydropower potential in China:378,000 MW, corresponding to an annual generation of $1{,}923\times10^{12}$W·h.

More details about the fifth or the latest survey are as follows.

## 2 Latest survey results

More than 11,000 hydropower sites were identified in the fifth survey. They were classified into four categories according to the amount of investigation and design work already made to each site.

(1) Hydropower sites in Category A include those sites with a considerable amount of

geological exploration work, including core drilling of the dam foundation, and engineering studies carried out, sufficient to the requirements of feasibility studies.

(2) Category B power sites include those with limited geological investigation and only general development conditions were assessed.

(3) Category C power sites include those that have only been reconnoitered by field visits and development concepts have been suggested, but no geological exploration was made.

(4) Category D power sites represent those that have only been studied in the office (on the maps), no field work has been done.

The total capacity of Category A and Category B power sites accounts for nearly one third of the total exploitable potential at the time when the survey was made.

The following table shows the distribution of hydropower potential in China according to river systems.

The following remarks can be added to Table 24.1.

(1) The Yangtze River is by far the most powerful river in China. A large portion of the hydropower potential of the Yangtze River is in its upper reaches and major tributaries.

(2) The Yellow River, although ranked second largest in drainage area, possesses only 6.1% of the total potential. Since the annual runoff of the Yellow River is only 1/20 of that of the Yangtze.

Table 24.1 Hydro-potential in China (according to river systems)

| Rivers | Theoretical potential | Exploitable potential | | |
|---|---|---|---|---|
| | 1000MW | 1000MW | $\times 10^{12}$W·h | % |
| Yangtze | 268.02 | 197.24 | 1,027.498 | 53.4 |
| Yellow | 40.55 | 28.00 | 116.991 | 6.1 |
| Pearl | 33.48 | 24.85 | 112.478 | 5.8 |
| Hai | 2.94 | 2.13 | 5.168 | 0.3 |
| Huai | 1.44 | 0.66 | 1.894 | 0.1 |
| Northeastern R. | 15.31 | 13.70 | 43.942 | 2.3 |
| Southeastern R. | 20.67 | 13.90 | 54.741 | 2.9 |
| Southwestern R. | 96.90 | 37.68 | 209.868 | 10.9 |
| Yalutsangpo | 159.74 | 50.38 | 296.858 | 15.4 |
| Inland Rivers | 36.99 | 9.97 | 53.866 | 2.8 |
| Total | 676.05 | 378.53 | 1,923.304 | 100.0 |

(3) The Hai River and Huai Rivers are important rivers for agricultural use, but possess little power potential.

(4) The Yalutsangpo River and the Southwestern Rivers which include the Lancan River (Upper Mekong) and the Nu River (Upper Salwen) possess a great amount of hydropower potential as they flow along or across the edge of the Ti-betan Plateau with a big discharge.

The distribution of hydropower potential according to regions is listed in Table 24.2.

The following remarks can be attached to Table 24.2.

(1) Two-thirds of the total exploitable hydropower potential is concentrated in China's

Southwest.

(2) The hydro potential in North China, Northeast and East China claims only a small percentage of the total, but the absolute values still represent a significant amount.

Table 24.2 Hydropower potential in China (according to regions)

| Region | Theoretical potential | Exploitable potential | | |
|---|---|---|---|---|
| | 1000MW | 1000MW | $\times 10^{12}$W·h | % |
| North China | 12.30 | 6.92 | 23.225 | 1.2 |
| Northeast | 12.12 | 11.99 | 38.391 | 2.0 |
| East China | 30.05 | 17.90 | 68.794 | 3.6 |
| Central South | 64.08 | 67.43 | 297.365 | 15.5 |
| Southwest | 473.31 | 232.34 | 1,305.036 | 67.8 |
| Northwest | 84.18 | 41.94 | 190.493 | 9.9 |
| Total | 676.05 | 378.53 | 1,923.304 | 100.0 |

(3) In Central South region, the exploitable potential is somewhat larger than the theoretical potential. This is due to the fact that the huge Three Gorges project is situated in Hubei Province of the Central South region, while a part of the potential utilized by this project belongs theoretically to Sichuan Province of the Southwest region.

The hydropower potential in Taiwan Province is not included in the current survey.

## New Words and Expressions

1.survey *vt.*, *n.* 调查，测量，综述，概括的研究，测量图，述评

2.potential *a.* 潜在的，可能的；势（差）的，位（差）的，有势的，无旋的，电位的；*n.* 潜力（能）；（动力）资源，蕴藏量；势能，位能；位（势）函数

3.supplement *vt.* 补（充，足，加，遗），增补[求]，添[追]加；*n.* 补（充，编，遗），增补[刊]，副刊，附录，补角

4.esteem *vt.* 尊[珍]重，重视，认为，看作；*n.* 尊重[敬]，好评

5.outmatch *vt.* 胜[强，超]过，优于

6.deliberately *ad.* 谨慎地，故意地，预有准备地

7.territory *n.* 领土（地），土地，领域，范围

8.reconnoiter (or reconnoiter) *vt.* 踏[查]勘，勘测[查]，侦察

# Lesson 25 Differential Leveling

## 1 Definitions

*Differential leveling* is the operation of determining the elevations of points some distance apart. Usually this is accomplished by direct leveling. Differential leveling requires a series of setups of the instrument along the general route and, for each setup, a rod reading back to a point of known elevation and forward to a point of unknown elevation.

**A *bench mark* (B.M.)** is a definite point on an object of more or less permanent character, the elevation and location of which are known. Bench marks serve as points of reference for levels in a given locality. Their elevations are established by differential leveling. Permanent bench marks have been established throughout the United States by the U. S. Geological Survey and the U. S. Coast and Geodetic Survey; these consist of bronze plates set in stone or concrete and marked with the elevation above mean sea level. Other agencies have established permanent monuments. For any survey or construction enterprise, local bench marks are established, based on some selected datum and employing natural or artificial objects such as stones, pegs, spikes in trees or pavements, and marks painted or chiseled on street curbs.

In some areas, the elevation of bench marks may be altered by earth movements such as those caused by earthquakes, slides, lowering of water tables, pumping from oil fields, mining, or construction.

**A *turning point* (T.P.)** is an intervening point between two bench marks, upon which point foresight and backsight rod readings are taken. The nature of the turning point is usually indicated in the notes, but no record is made of its location unless it is to be reused. A bench mark may be used as a turning point.

**A *backsight* (B.S.)** is a rod reading taken on a point of known elevation. It is sometimes called a *plus sight.*

**A *foresight* (F.S.)** is a rod reading taken on a point the elevation of which is to be determined. It is sometimes called a *minus sight.*

**The *height of instrument* (H.I.)** is the elevation of the line of sight of the telescope when the instrument is leveled.

In surveying with the transit, the terms backsight, foresight, and height of instrument have meanings different from those here defined.

## 2 Procedure

In Fig. 25.1 B.M.$_1$ represents a point of known elevation (bench mark) and B.M.$_2$ represents a bench mark to be established some distance away. It is desired to determine the elevation of B.M.$_2$. The rod is held at B.M.$_1$, and the level is set up in some convenient location, as $L_1$ ,along the general route but not necessarily on the direct line joining B.M.$_1$ and B.M.$_2$ A backsight is taken on B.M.$_1$.

The rodman then goes forward and, as directed by the leveler, chooses a turning point T.P.$_1$ at some convenient spot within the range of the telescope along the general route B.M.$_1$ to B.M.$_2$. It is desirable, but not necessary, that each foresight distance, as $L_1$—T.P.$_1$, be made approximately equal to its corresponding backsight distance, as B.M.$_1$—$L_1$. The rod is held on the turning point, and a foresight is taken. The leveler then sets up the instrument at some favorable point, as $L_2$, and takes a backsight to the rod held on the turning point; the rodman goes forward to establish a second turning point T.P.$_2$; and so the process is repeated until finally a foresight is taken on the terminal point B.M.$_2$.

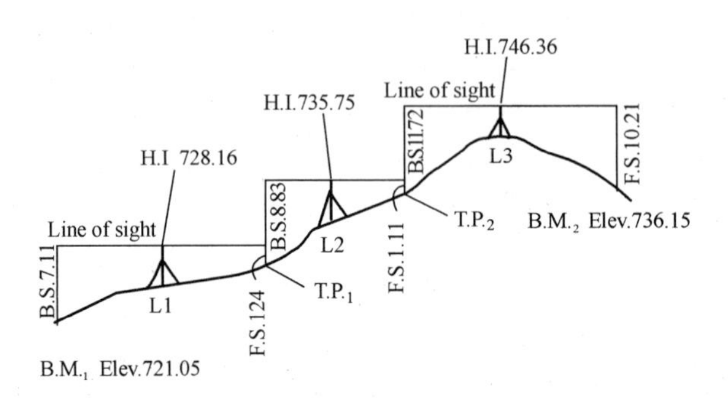


Fig. 25.1 Differential leveling

It is seen in Fig. 25.1 that a backsight added to the elevation of a point on which the backsight is taken gives the height of instrument, and that a foresight subtracted from the height of instrument determines the elevation of the point on which the foresight is taken. Also, the difference between the backsight taken on a given point and the foresight taken on the following point is equal to the difference in elevation between the two points. It follows that the difference between the sum of all backsight and the sum of all foresights gives the difference in elevation between the bench marks.

Sometimes in leveling for a tunnel or a building it is necessary to take rod readings on points which are at a higher elevation than that of the H.I. In such cases the rod is held inverted and in the field notes each such backsight is indicated with a minus sign and each foresight with a plus sign.

When several bench marks are to be established along a given route, each intermediate bench mark is made a turning point in the line of levels. Elevations of bench marks are checked usually by continuing the line of levels back to the initial bench mark. A line of levels that ends at the point of beginning is called a level circuit. If the circuit checks within the prescribed limits of error, it is considered that the elevations of all turning points are correct within prescribed limits, but such closure does not serve as a check on side shots taken from the circuit.

### 3 Balancing Backsight and Foresight Distances

If a foresight distance were equal to the corresponding backsight distance, any error in readings due to earth's curvature and to atmospheric refraction (under uniform conditions) would be eliminated.

In ordinary leveling no special attempt is made to balance each foresight distance against the preceding backsight distance. Whether or not such distances are approximately balanced between bench marks will depend upon the desired precision. The effect of earth's curvature and atmospheric refraction is slight unless there is an abnormal difference between backsight and foresight distances. The effect of instrumental errors is likely to be of considerably greater consequence with regard to the balancing of these distances. The chances are that there is not absolute parallelism between the line of sight and the axis of the level tube. The error in a rod reading due to this imperfection of adjustment is proportional to the distance from the instrument to the rod and is of the same sign for a backsight as for a foresight. Since backsights are added and foresights are subtracted, this instrumental error is eliminated if, between bench marks, the sum of the foresight distances is made equal to the sum of the backsight distances.

In leveling of ordinary precision, with an instrument in good adjustment, the backsight and foresight distances are not measured, and no special attempt is made to equalize them even by estimation. Normally they tend to balance in the long run. However, a succession of very long backsight distances and very short corresponding foresight distances, or vice versa, as would be likely for leveling between two points having a large difference in elevation, would produce a systematic error of appreciable magnitude.

For leveling of moderately high precision it is necessary to equalize backsight and foresight distances between bench marks. In less refined leveling, distances are usually determined by pacing; in precise leveling, they are usually measured with the stadia or the gradienter.

In leveling uphill or downhill a balance between foresight and backsight distances can be obtained with a minimum number of setups by following a zigzag course.

The effect of earth's curvature and atmospheric refraction cannot be entirely eliminated by making the sum of the foresight distances equal to that of the backsight distances; rather it would be necessary that each foresight distance be made equal to the corresponding backsight distance.

## New Words and Expressions

1.differential leveling 水准测量
2.leveling *n.* 水准测量
3.bench mark (B.M.) 水准点
4.more or less 大约；或多或少
5.rodman *n.* 司尺员
6.leveler *n.* 观测员
7.it follows that 由此可见
8.side shot 旁测点
9.earth's curvature 地球曲率
10.atmostpheric refraction 大气折光
11.be of consequence 有意义，重要
12.with regard to 就…而论
13.the chances are that 有可能
14.pacing *n.* 步测
15.stadia *n.* 视距尺；视距测量
16.gradienter *n.* 倾斜测定仪
17.zigzag *n.* 之字形，折线
18.monument *n.* 纪念物；记录
19.enterprise *n.* 企业；计划
20.datum *n.* 基准面；基点；数据
21.spike *n.* 道钉
22.peg *n.* 桩；钉；测标
23.curb *n.* 路沿石，路牙
24.chisel *v.*，*n.* 雕，凿，刻
25.water table 地下水位
26.turning point (T.P.) 转点，转折点

27.rod reading　水准尺读数

28.foresight　*n.* 前视

29.backsight　*n.* 后视

30.elevation　*n.* 高程；标高；海拔

31.level　*vt.* 调平

32.transit　*n.* 经纬

## Reading Material　Fundamentals of CAD

CAD is an acronym. Some people use it for 'computer-aided drawing', others for 'computer-aided design'. They seem to agree on 'computer-aided', and indeed these words are a useful reminder that the computer is merely a device which exists to assist us. This puts the machine in its rightful place, for we humans ought to remain in full control. It is over the third word that confusion sometimes arises.

There are now many CAD systems available to design organizations. They vary in cost, scope, capability, and in suitability for the work of the particular office. All but the simplest systems are intended to be rather more than aids to drawing production. However none is capable of aiding all aspects of the design process. In reality, all systems fall somewhere between "computer-aided drawing" and "computer-aided design". So "CAD" will be used just as convenient label for a complex technique.

Many people find that technical jargon is a barrier to their comprehension of anything connected with computers. However it is difficult to avoid its use entirely. The aim here will be to limit it as much as possible, and to explain terms when first used. We have already met the problem with "CAD" itself, and have met the widely used word 'system'. The latter will specifically mean the combination of several items of equipment which are linked together to form the computer configuration, plus the programs or instructions which enable the equipment to function under the user's control. So a "system" is "hardware" plus "software" the equipment and the programs.

Where there is a history of using computers particularly within an engineering design office, it has mainly been for elaborate calculation such as structural analyses, and for administrative chores. Such tasks are essentially operations on numbers, and this is work for which computers are ideally suited.

However designers tend to communicate most project information in graphical form. Drawing have always been central to the whole construction process, and are likely to continue as such in the foreseeable future.

It has taken some time to teach computers to cope with information in graphical form. But the incentives are evident. A design office might typically spend only 5% or 10% of its collective effort on calculations. However the production of drawings might well represent 40% or more of its total workload. So any improvement that can be achieved in drawing efficiency ought to have a high impact. This undoubtedly accounts for much of the huge current interest in CAD.

Although this interest seems very recent, the origins of the technique in fact date back to around 1960. Much progress was made during the 1960s and early 1970s. There was impetus from the

American space programme, and from the aerospace, automotive and electronics industries.

In the 1970's, an examination of the commercially available systems might well have indicated that they were still too expensive and too lacking in practical features to be attractive for normal design office use. Nevertheless the race was certainly on among potential suppliers for a share of a big market that was undoubtedly coming.

Automatically plotted drawings produced as a by-product of an engineering applications program are no recent achievement. As an example, if we examine highway design, we find that in the late 1960s and early 1970s the geometry of most important new roads was being drawn by computer. Plans, sections and perspective views of roads could be produced automatically. However the particular programs used were limited in scope to roads only, and so were special-purpose applications programs. Moreover they worked best for long roads in rural or, better still, desert areas, They worked least well for complex road interchanges in the urban environment. Although they remain immensely valuable, such programs would not be classified as CAD systems because they lack a capability for interactive graphics. They do not permit the designer to carry on a dialogue with the computer to gradually build up, refine and modify his design information.

In building design, a similar approach of generating drawings automatically as the by-product of design programs would be less successful. It would be possible at all only for highly coordinated system buildings.

Therefore the widespread use of computer techniques has had to await the introduction of the interactive graphics systems now labelled as CAD systems. These permit the designer /draughtsman to sit at a workstation and to issue a command and other information to the computer. The computer then performs some function as instructed and displays the result on a graphics screen for the user to check. Assuming all is well, the user can proceed to another step in the process, and another, until the whole design task is completed.

Thus the user interacts with the computer. There is teamwork between him and the system. This teamwork is usually successful because the computer is adept at repetitious and mundane tasks, has a large and extremely accurate memory, can perform systematic tasks very rapidly, and does not get tired or even bored. The human is much superior at controlling the process or steering. He does this by exercising human qualities including his judgement, experience, intuition, imagination and intelligence.

## New Words and Expressions

1.acronym *n.*（首字母）缩写词，缩语，简称

2.hardware *n.* 硬件，计算机部件

3.elaborate *a.* 精心的，费事的，复杂的

4.chore *n.* 零星工作，零活，杂物

5.foreseeable *a.* 可预见的，有远见的

6.in the foreseeable future 在可预见的未来一段时间内

7.incentive *a.*刺激的，鼓励的；*n.*刺激，鼓励，动机

8.impetus *n.* 动力，刺激，促进，推动

9.interchange *v.* 交换，交替，*n.* 交换，道路互通式立体交叉

10.immensely *ad.* 广大地，无限地，非常，很

11.interactive *a.* 相互作用，交互的，人—机对话的

12.draughtsman *n.* 绘图员，制图员，起草员

13.repetitious *a.* 重复的，反复的

14.mundane *a.* 世间的，世俗的

15.jargon *n.* 行话，术语，难懂的话

# Lesson 26  Construction and Equipment

Many of the major contractors, particularly in the west, started in business building dams and canals for the U. S. Bureau of Reclamation. The work was usually bid on unit prices and the quantities were listed as estimated by the owner's engineer. The contract was let on the basis of the lowest bid computed on the total price for all units bid upon.

The Bureau of Reclamation maintained its own construction force until the middle 1920's and preferred to do its own work "force account". The "force account" work was performed by a rather large and highly efficient construction organization within the Bureau and was headed up by several outstanding "construction superintendents" reporting to the chief engineer. They worked on plans and specifications prepared by The Bureau and the work was inspected by Bureau engineer. They requisitioned equipment and, with the chief engineer's approval, hired their foremen, master, mechanics and workmen, and ran their own cook houses and camps. Careful costs were maintained and high standards of work were reached.

As the policy shifted, from building major dams by force account and contracting smaller structures, to calling for bids on all work and awarding contracts when the contractor's figures were at or below the "force account" bid, the force account organization was then gradually disbanded.

The dam builders were the first to practice from sheer necessity, the charting and scheduling of sequence and timing of all major items of work. Materials and equipment were furnished to meet a preplanned schedule. From the late 1920's this was done laboriously by hand, until the advent of the computer, when the "critical path method" was introduced in 1958. The CPM required that a project be broken down into component activities which could be presented in the form of network diagram showing their sequential relationships to one another.

A critical path of the construction operation at Itaipu, as an example, had run through the following major operations in sequence or concurrently: ①An assured supply of cement; ②Award of contract and adequate move in time allowance; ③Construction of camp facilities and haul road; ④Excavation of diversion channel; ⑤Construction of cofferdams; ⑥Furnishing and erection of the concrete-placing; ⑦Placement of concrete in dam and powerhouse to EL 139 and installation of intake gate; ⑧Procurement and installation of turbines and generators.

Prior to 1930, the construction equipment was all small. The jobs were of fairly long duration and the contractor was always short of funds. As a consequence in the late 1920's the "joint venture" was conceived. New, larger, and more productive equipment continued to be developed. As an example, up to the early 1960's the most popular loading machine for rock was the power shovel. At that time, development of the front-end loader accelerated and in the 1970's the front-end loader has displaced the power shovel in most applications. But today, the 15-cu-yd electric shovel has been developed and is being used in Itaipu construction site.

When considering all types of dams, the requirements for equipment will be varied that selection of equipment will depend on the individual job. In the case of embankment dam, the major equipment generally used includes excavating, loading, transporting and compacting machinery; such as track-type pneumatic drills, stationary compressed-air plant and air-delivery system, face shovel, backhoe, dragline, scraper (singly or in tandem), tracked tractor, bottom dump trailer, rear dump truck (off-highway), bulldozer, grader, both sheep-foot and rubber-tired rollers, vibratory roller, etc.

In the case of concrete dam, much of the equipment used in excavating and loading can be salvaged after the excavations are completed. Whereas the concrete system including plant for aggregate preparation, stockpiling, batching and mixing facilities, concrete-delivery and concrete-placing equipment must be installed. The conveyor belts, which have the advantage of continuous operation, are chiefly used in the processing of aggregate and transporting wet concrete to the placement. The most reliable and widely used equipment for placing concrete are cableway or rail-mounted whirling gantry crane, concrete bucket and vibrator, etc.

No discussion of construction equipment would be complete without considering the function of the Manufacturer's Representative or Sales Engineer. Nowadays the MR or Sales Engineer is usually a graduate engineer who has specialized in the capabilities of equipment produced by one manufacturer. It is a mistake to regard these specialists as just salesmen or "peddlers".

## New Words and Expressions

1.contract *v.* 订立合同（契约），缔结；*n.*合同，契约，承包，合同

2.contractor *n.* 承包者（人，商），合同户，包工（头），单位

3.bid *n.*，*v.* 出价，投标，喊价，努力，尝试

4.owner *n.* 所有者，物主，业主

5.force account 计工制

6.superintendent *n.* 监造师，总工程师，监督人，管理人

7.chief engineer 总工程师

8.owner's engineer 甲方代表，业主工程师

9.preplanned schedule 预定进度表

10.network *n.* 网络，网格

11.move-in 进点

12.haul *n.* 搬运，运程（距），运输量

13.cofferdam *n.* 围堰，防水堰，*v.*修筑围堰

14.trestle *n.* 栈桥，支架，旱桥

15.procurement *n.* 征（收，采）购，取（获）得

16.approval *v.* 赞成，同意，认可，批准

17.foreman *n.* 领班，工长

18.mechanic *n.* 机（修，械）工（人），技工，机械师（员）

19.camp *n.* 帐篷，（野，设）营（地），工房

20.call for bid 招标

21.awarding contract 签订合同

22.charting *n.* 制图（表），填（绘）图

23.scheduling *n.* 制进度表，编制计划，调度，计划

24.job *n.* 工程，任务

25.joint venture 联营企业

26.loading machine 装料机

27.power shovel 动力铲

28.front-end loader 前端装料机

29.track-type 履带式

30.pneumatic *a.* 气（风）动的，风力的，有空气的

31drill *n.* 钻

32.stationary *a.* 不动的，静止的，固定的，平稳的

33.compressed-air plant 压缩空气装置
34.air-delivery system 输气系统
35.face shovel 正（向）铲
36.backhoe *n.* 反向铲
37.dragline *n.* 绳斗（电）铲，拉（索）铲挖土机
38.scraper *n.* 铲运机
39.tandem *a.* 串联的，前后排列的；
in~ 一前一后地，相互合作地，协力地
40.botton dump 底卸式
41.trailer *n.* 拖车
42.rear dump 后卸式
43.off-highway 越野式的
44.bulldozer *n.* 推土机，压路机
45.grader *n.* 平地（土，路）机，推土机
46.sheep-foot roller 羊脚碾
47.rubber-tired roller 轮胎碾
48.vibratory roller 振动碾
49.salvage *n.* 废物利用，救难费
50.stockpiling *n.* 装堆，存料，存货
51.batching *n.* 配料，计量
52.conveyor *n.* 输送机；~belt 运输带，皮带运输机
53.wet concrete 流态混凝土
54.cableway *n.* 索道，缆道
55.rail-mounted 有轨的，装有轨道的
56.whirling gantry crane 门式旋转起重机
57.concrete bucket 混凝土吊罐
58.vibrator 振动器
59.sales engineer 销售工程师
60.graduate engineer 有学位的工程师

# Reading Material Concrete Construction

## 1 Placing Concrete

If concrete is placed on earth, the earth should be moistened sufficiently to prevent it from robbing the concrete of its water. If fresh concrete is to be placed on or adjacent to concrete that has set, the surface of the old concrete should be cleaned thoroughly, preferably with a high-pressure air and water jet and steel-wire brushes. The surface should be wet, but there should be no standing water. A small quantity of cement grout should be brushed over the entire area, then followed immediately with the application of a 1/2-in. layer of mortar. The fresh concrete should be placed on or against the mortar.

In order to reduce the segregation resulting from movement after it is placed, concrete should be placed as nearly as practicable in its final location. It should be placed in layers whose thickness will permit uniform compaction. The time lapse between the placing of layers should be limited to assure perfect bond between the fresh and previously placed concrete.

In placing concrete in deep forms, a tremie should be used to limit the free fall to not over 3 or 4 ft, in order to prevent segregation. A tremie is a pipe made of lightweight metal, having adjustable lengths and attached to the bottom of a hopper into which the concrete is deposited. As the forms are filled, sections of the pipe may be removed.

Immediately after concrete is placed, it should be compacted by hand puddling or a mechanical vibrator to eliminate voids. The vibrator should be left in one position only long enough to reduce the concrete around it to a plastic mass; then the vibrator should be moved, or segregation of the aggregate will occur. In general, the vibrator should not be permitted to penetrate concrete in the prior lift.

The primary advantage of vibrating is that it permits the use of a drier concrete, which has a higher strength because of the reduced water content. Among the advantages of vibrating concrete are the following.

(1) The reduced water permits a reduction in the cement and fine aggregate because less cement paste is needed.

(2) The lower water content reduces shrinkage and voids.

(3) The drier concrete reduces the cost of finishing the surface.

(4) Mechanical vibration can replace three to eight hand puddlers.

(5) The lower water content increases the strength of the concrete.

(6) The drier mix permits the removal of some forms more quickly, which may reduce the cost of forms.

## 2 Curing Concrete

If concrete is to attain its maximum strength and other desirable properties, it should be cured with adequate moisture and at a favorable temperature. Failure to provide these conditions may result in an inferior concrete.

The initial moisture in concrete is adequate to hydrate all the cement, provided it is not permitted to evaporate before it is used. Curing should prevent the loss of initial moisture, or it should replace the moisture that does evaporate. This may be accomplished by several methods, such as leaving the forms in place, keeping the surface wet, or covering the surface with a liquid curing compound, which forms a watertight membrane that prevents the escape of the initial water. Curing compounds may be applied by brushes or pressure sprayers. A gallon will cover 200 to 300 sq ft.

Concrete should be placed at a temperature not less than 40 or more than 80 F. A lower temperature will reduce the rate of setting, while a higher temperature will reduce the ultimate strength.

## 3 Placing Concrete in Cold Weather

When concrete is placed during cold weather, it usually is necessary to preheat the water, the aggregate, or both in order that the initial temperature will assure an early set and gain in strength. Preheating the water is the most effective method of providing the required temperature. For this purpose a water is the most effective method of providing the required temperature. For this purpose a water reservoir should be equipped with pipe coils through which steam can be passed, or steam may be discharged directly into the water, several outlets being used to given better distribution of the heat.

When the temperatures of the ingredients are known, some specific charts may be used to determine the temperature of concrete. A straight line across all three scales, passing through any two known temperature, will permit the determination of the third temperature. If the sand is surface-dry, the solid lines of the scales giving the temperature of concrete should be used. However, if the sand contains about 3 percent moisture, the dotted lines should be used.

Specifications frequently require that freshly placed concrete shall be maintained at a temperature of not less than 70° F for 3 days or 50° F for 5 days after it is placed. Some suitable method must be provided to maintain the required temperature when cold weather is anticipated.

## New Words and Expressions

1.place *v.* 浇筑
2.rob *v.* 夺取
3.grout *n.* 薄浆，灰浆，水泥浆
4.segregation *n.* 离析
5.hopper *n.* 漏斗，加料斗
6.deposit *v.* 存放，储存
7.puddling *n.* 捣实
8.puddler *n.* 捣实器
9.cure *v.* 养护
10.inferior *a.* 较差的
11.sprayer *n.* 喷雾器，喷射装置
12.coil *n.* 环绕
13.ingredient *n.*（混合物的）组成部分
14.rob M of N 使 M 失去 N，从 M 夺取 N
15.cement paste 水泥浆

# Lesson 27 Engineering Economy in Water Resources Planning

Water resources planning involves many choices among physically feasible alternatives. Generally speaking, each choice among alternatives should be made on economic grounds. Each alternative that is given serious consideration should be expressed in money units, before the choice is made. In fact, unless alternatives can be expressed in money units, the items involved in such choices are incommensurable. For instance, the money unit is the only measuring unit that can be applied to such diverse items as steel pipe, kilowatt-hours of electricity, hours of skilled and unskilled labor, and reduction of the hazard of flood damages.

Writers on engineering economy often quote the classic questions that Gen. John J.Carth applied to engineering proposals when he was chief engineer of the New York Telephone Company. These questions were why do this at all? Why do it now? Why do it this way? These may be viewed as different aspects of the general question, will it pay?

As these questions suggest, most engineering proposals involve major sets of alternatives, with many subordinate sets of alternative involved in each major alternative. For example, consider the question of whether to use hydro or thermal power to add needed capacity to an electric-power system. For the hydro alternative, a number of different possible sites may be available. At a particular site, there may be different possible designs for a diversion dam, different type and sizes of conduit to carry water from the point of diversion to the power plant, different possibilities for number, type, and size of turbines at the power plant, and so on. Each sub-alternative is likely to have its own sub-alternatives. At every stage in the analysis, it is necessary to consider relative economy in order to make a rational choice among the possible alternatives and thus to establish the most favorable overall plan.

It is helpful to think of an economy study as involving the following steps:

(1) Each alternative that seems promising should be identified and clearly defined in physical terms.

(2) Insofar as practicable, the physical estimates for each alternative should be translated into money estimates. Generally speaking, money estimates should be made of those receipts and disbursements which will be influenced by the choice among the alternatives. Estimates should be made of the dates as well as the magnitudes of the receipts and disbursements. This requires estimates of the lives and salvage values, if any, of the structures and other assets required for each alternative. It also calls for a decision regarding the length of the study period, the period of time for which the economy study is to be made.

(3) Usually the money estimates need to be placed on a comparable basis by appropriate

conversions that make use of the mathematics of compound interest. These conversions should use as an interest rate the minimum attractive rate of return that is appropriate in the particular circumstances.

(4) A choice (or recommendation for a choice) among the alternatives must be made. This choice is properly influenced both by the comparison in terms of money units and by other matters that it has not been practicable to reduce to money terms.

A simple example of an annual cost comparison is listed below:

Two alternative plans are considered for a section of an aqueduct. Plan A uses a tunnel, plan B uses a section of lined canal and a section of steel flume. In plan A, the estimated first cost of the tunnel is $450,000, its estimated annual maintenance cost is $4,000, and its estimated life is 100yr. Estimated first costs and lives for the elements of plan B are canal (not including lining), $120,000, 100yr; canal lining, $50,000, 20 yr; flume, $90,000, 50 yr. The annual maintenance cost is $10,500. The interest rate to be used in the economy study is 6 percent per annum. The study period is 100 yr. All salvage values are assumed to be negligible. There are no estimated revenue differences between the two plans, and no other cost differences are anticipated, (For instance, there is no expected difference in water loss between the two alternatives.)

The comparison of equivalent annual costs for the two plans is as follows:

Plan A:

Capital recovery cost for tunnel: $450,000×0.06018=$27,081

Annual maintenance cost: $4,000

Total annual cost: $31,081

Plan B:

Capital recovery cost for canal: $120,000×0.06018=$7,222

Capital recovery cost for canal lining: $50,000×0.08718=4,359

Capital recovery cost for flume: $90,000×0.06344=5,710

Annual maintenance cost: $10,500

Total annual cost: $27,791

In the preceding tabulation, the first-cost figures for tunnel, canal, lining, and flume are multiplied by compound-interest factors that depend on their respective estimated lives and on the 6 percent interest rate. The appropriate factor to convert an investment into an equivalent annual cost is designated as the capital recovery factor and may be computed from the expression $i(1+i)^n$ $[(1+i)^n-1]$ where $i$ represents the interest rate per annum (expressed as a decimal fraction) and $n$ represents the years of estimated life.

When any present sum of money is multiplied by the capital recovery factory for $n$ year and interest rate $i$, the product is an annual figure sufficient to repay exactly the present sum in $n$ yr with interest rate $i$. For example, the tabulated costs in plan B show $4,359 as the capital recovery cost of $50,000, including 6 percent interest on each year's unpaid balance.

The comparison of the $31,081 annual cost of plan A with the $27,791 annual cost of plan B is representative of the innumerable annual cost comparisons that may be made in engineering economy studies. It is common for alternative designs to require different investments in fixed assets. In this example, the total investment figures are \$450,000 and \$200,000, respectively. The extra investment of \$190,000 produces advantages that have a money value, such as lower maintenance costs and longer lives for the assets. In this instance, the annual cost comparison tells us that these advantages are insufficient to justify the extra investment.

Plan B therefore should be selected unless plan A has some other advantages.

How to select an interest rate for an economy study?

The conclusions from a comparison of annual costs obviously are greatly influenced by the interest rate used in computing the costs. The choice of the interest rate to be used in economy studies to guide the design of an engineering project will have considerable influence on the design selected. With a very low rate, many proposed investments (such as the tunnel of plan A) will appear to be economical even though the same investments would seem unduly costly with an interest rate of 6 or 8 percent. In effect, the decision to use a particular rate, such as the 6 percent that was used in comparing plans A and B, is a decision that the rate selected is the minimum attractive rate of return.

In private enterprise, the interest rate used in economy studies should ordinarily be not less than a figure that reflects the overall cost of capital to the enterprise equity capital as well as borrowed capital.

The common practice in economy studies for public works in the United States has been to use an interest rate equal to the bare cost of borrowed money for the public body in question. Such rates differ from time to time and from one public body to another. The rate used in 1978 for federal river-basin projects was 5.38 percent.

How to estimate lives of hydraulic structures?

A bulletin of the Internal Revenue Service gives estimated average lives for many thousands of different types of industrial assets. The lives (in years) given for certain elements of hydraulic projects are listed in Table 27.1.

Such countrywide estimates of average lives may be helpful even though they are not necessarily the most appropriate figures to use in any given instance. Moreover, a conservative life estimate to use in an economy study usually should be shorter than the full expected service life. Many factors combine to cause economic lives to be shorter than full service lives. A competitive private industry might use a 10-yr payoff period for assets having expected lives of 20-yr or more. Costs of public river-basin projects are usually computed on the basis of a 100-yr life.

For long-lived assets, a large difference in estimated life has less effect on annual cost than a moderate difference in interest rate. For example, assume a given life estimate is increased from 35 to 100 yr and at the same time the interest rate used is increased from 5 to 6 percent, the increases of annual cost due to the higher interest rate is greater than the reduction of cost due to the estimate of longer life.

Table 27.1 Lives (in years) for elements of hydraulic projects

| | | | |
|---|---|---|---|
| Barges | 12 | Pipes: | |
| Booms log | 15 | Cast-iron | |
| Canals and ditches | 75 | 2~4 in | 50 |
| Coagulating basins | 50 | 4~6 in | 65 |
| Construction equipment | 5 | 8~10 in | 75 |
| Dams: | | 12 in and over | 100 |
| Crib | 25 | Concrete | 20 |
| Earthen, concrete, or masonry | 150 | Stell: | |
| Loose rock | 60 | Under 4 in | 30 |
| Steel | 40 | Over 4 in | 40 |
| Filters | 50 | Transite 6 in | 50 |
| Flumes: | | Transmission lines | 30 |
| Concrete or masonry | 75 | Tugs | 13 |
| Steel | 50 | Wood-stave: | |
| Wood | 25 | 14 in and larger | 33 |
| Fossil-fuel power | 28 | 3~12 in | 20 |
| Generators: | | Pumps | 18~25 |
| Above 3,000 kW | 28 | Reservoirs | 75 |
| 1,000~3,000 kW | 25 | Standpipes | 50 |
| 500~1,000 kW | 17~25 | Tanks: | |
| Below 500 | 14~17 | Concrete | 50 |
| Hydrants | 50 | Steel | 40 |
| Marine construction equipment | 12 | Wood | 20 |
| Meters-water | 30 | Tunnels | 100 |
| Nuclear power plants | 20 | Turbines-hydraulic | 35 |
| Penstocks | 50 | Wells | 40~50 |

**(to be continued)**

# New Words and Expressions

1.commensurable *a.* 可公度的；~with 相应的；~to 相当的；un- 不一

2.hazard *n.* 危险，公害，*vt.* 使遭危险

3.quot *v.* 引用，引述，开（价）

4.subordinate *a.*下级的，次要的；*n.* 下属；*vt.* 服从

5.conduit *n.* 管道，水道（管）

6.insofar *ad.* 到这个程序（范围）；-as 在…限度内

7.receipt *n.* 收到收据，*vt.* 出…的收据

8.disburse *vt.* 支付，支出；*n.* -ment

9.salvage *n.* 打捞，待报废器材，换算

10.asset *n.* 财富，(*pl.*)资产

11.conversion *n.* 转化，变换，换算

12.recommendation *n.* 推荐，介绍

13.aqueduct *n.* 沟渠，输水道

14.tunnel *n.* 地道，隧洞

15.annum *n.* 一年

16.repay *v.* 付还，偿还，报答；–ment *n.*

17.unduly *ad.* 过度（分）地，不适当地

18.payoff *n.* 偿清，支付，收效，发工资

19.equity *n.* 公平，正当，无固定利息的股票；-capital 新企业投资

20.compound interest 复利；~factor 复利因子

21.the minimum attractive rate of return 最小回收利率

22.Capital recovery cost 资金偿还(恢复)贷
23.first-cost figure 初始投资数字
24.decimal fraction （十进制）小数
25.present value (worth) 现值
26.present sum 现值总量
27.bare cost 裸值
28.servic life 服务寿命
29.long-lived assets 长寿命的固定资产
30.in question 上述那个，讨论中的

## Reading Material Engineering Economy in Water-Resources planning

**(continue)**

The average annual damage from floods in a river basin is estimated to be $400,000. Estimates are made for several alternate proposals for flood-mitigation works. Channel improvements would increase the capacity of the stream to carry flood discharge. There are two possible sites, A and B, for a dam and storage reservoir. Because dam site A is located in the reservoir area for B, one or the other of these sites may be used, but not both. Either site may be used alone or may be combined with channel improvement. It is also possible to use channel improvement alone. Table 27.2 shows the estimated first cost of each project, the estimated annual damages due to floods with each project, the annual investment charges, using an interest rate of 3 percent, and the estimated annual disbursements for operation and maintenance. The life of the channel improvements is estimated as 25 yr.; the life of the dam and reservoir is estimated as 100 yr. The final column of the table gives the sum of annual damages and annual project costs. This sum is a minimum for project Ⅲ, the development at site B alone.

The more conventional way to analyze such public works proposals is by means of the benefit-cost ratio. Table 27.2 illustrates such an analysis. The benefits (to the flood sufferers) are the estimated annual reductions in flood damages. The costs (to the government and therefore to the taxpayers) are the annual investment charges plus annual disbursements for operation and maintenance.

Table 27.2 Economic analysis of proposals for flood mitigation

| Project | Investment ($) | Average annual flood damage ($) | Annual investment charges ($) | Annual operation and maintenance ($) | Sum of annual damages and project costs ($) |
|---|---|---|---|---|---|
| No flood mitigation at all | 0 | 400,000 | 0 | 0 | 400,000 |
| Ⅰ.Channel improvement alone | 500,000 | 250,000 | 28,720 | 100,000 | 378,720 |
| Ⅱ.Development at site A alone | 3,000,000 | 190,000 | 94,950 | 60,000 | 344,950 |
| Ⅲ. Development at site B alone | 4,000,000 | 125,000 | 126,600 | 80,000 | 334,600 |
| Ⅳ.Site A with channel improvement | 3,500,000 | 100,000 | 123,670 | 160,000 | 383,670 |
| Ⅴ.Site B with channel improvement | 4,500,000 | 60,000 | 155,320 | 180,000 | 395,320 |

The five benefit-cost ratios of Table 27.2 do not in themselves provide enough information to make an economic choice among the five projects. To use the benefit-cost ratio as a sound basis for project formulation, additional calculations are necessary. The additional benefits added by each separable increment of costs should be computed, and the ratios of the increments of benefits to the corresponding increments of costs should be determined. Such an analysis will lead to selection of project Ⅲ, just as in the analysis shown in Table 27.3. The preceding statement assumes that extra costs are justifiable whenever the resulting benefits exceed the extra costs but are not justified if the resulting benefits are less than the extra costs. In other words, the most economical design is the one that gives the greatest excess of benefits over costs.

Table 27.3 Benefit-cost analysis of flood mitigation proposals

| Project | Annual benefits ($) | Annual costs ($) | Benefit-cost ratio | Benefit minus costs ($) |
|---|---|---|---|---|
| Ⅰ | 150,000 | 128,720 | 1.17 | 21,280 |
| Ⅱ | 210,000 | 154,950 | 1.36 | 55,050 |
| Ⅲ | 275,000 | 206,600 | 1.33 | 68,400 |
| Ⅳ | 300,000 | 283,670 | 1.06 | 16,330 |
| Ⅴ | 340,000 | 335,320 | 1.0 | 4,680 |

Thus project Ⅱ adds benefits of $60,000 over project Ⅰ, whereas costs are increased by only $26,230, the ratio of extra benefits to extra costs is 2.29. Similarly, the extra $51,650 of costs of project Ⅲ over project Ⅱ are justified by increased benefits of $65,000, the incremental benefit-cost ratio is 1.26. But projects Ⅳ and Ⅴ are clearly uneconomical as compared with project Ⅲ because the added benefits are considerably less than the extra costs required to produce the benefits.

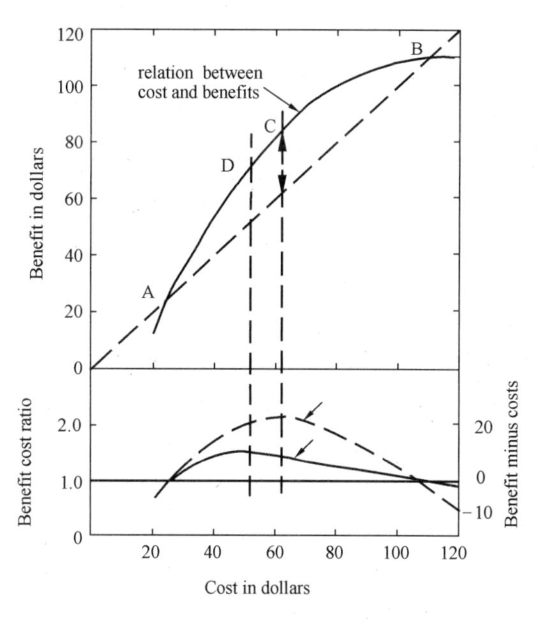


Fig. 27.1 Relation between costs and benefits for a hypothetical project

Fig. 27.1 shows the benefits and costs of a hypothetical project graphically. From point A to B, benefits exceed costs, i. e., the benefit-cost ratio exceeds one. The curve of benefits minus costs (i. e., net benefits) shows a maximum at C. Beyond this point, each dollar of costs returns less than a dollar of benefits, that is, $\Delta$ benefits/ $\Delta$ costs<1. The maximum ratio of benefits to costs occurs at D, and this should be the limit of project size if is desired to obtain a maximum rate of return on the investment. The increment from D to C is, however, economic since the rate of return on the increment exceeds the minimum attractive rate of return.

As suggested in the previous paragraph, a project may be evaluated in terms of rate of return. Using the estimated cost and benefit stream over the period of the project life, one may determine the rate of return on investment represented by the excess of benefits over costs. Projects may be ranked in merit on rate of return, and the decision to proceed can be based on a minimum acceptable rate of return as well as on a benefit-cost ratio.

Because a multiple-purpose project serves several different groups of beneficiaries, it is often necessary to allocate the cost among the several uses to fix the prices of water or power or to determine the contribution required of flood-mitigation beneficiaries.

There is no really satisfactory method of cost allocation in the sense that such a method would be equally applicable to all projects and would yield allocations which are unquestionably correct. Any method of allocation must first set aside the separable costs which are clearly chargeable to a single project function, such as the cost of the powerhouse, navigation locks, or fish ladders. The separable costs for a single function are usually estimated as the total project cost less the estimated cost with that function omitted. The real problem in cost allocation is the division of the joint costs (total cost less the sum of the separable costs) among the project functions. The two methods which seem most applicable to a multiple-purpose water project are the remaining-benefits method and the alternative justifiable-expenditure method. The application of these methods is illustrated by Table 27.4. The first line of the table presents the separable costs, which total \$1,180,000. Since the total cost to be allocated is \$1,765,000, the joint cost is \$585,000.

In the remaining-benefits method, the joint costs are assumed to be distributed in accordance with the differences between the separable costs (line 1) and the estimated benefits of each function (line 2). In no case, however, are the benefits assumed to be greater than the cost of an alternate single-purpose project which would provide equivalent benefits. Thus the remaining benefits (line 5) are the differences between the lesser value of lines 2 or 3 and the separable costs in line 1. The joint costs are distributed in proportion to these remaining benefits and added to the corresponding separable costs to obtain the total allocation(line 7).

Under the alternative justifiable-expenditure method the joint costs are assumed to be distributed in accordance with the differences between the separable costs and the estimated cost of a single-purpose project which would provide equivalent services and would itself be economically justifiable. These costs (line 8) are the differences between lines 1 and 3. The distributed joint costs (line 9) are added to the separable costs to obtain the total allocated costs (line 10). It should be noted that a large percentage difference in the allocations of joint costs is only a moderately small percentage difference in the allocated total cost because the separable costs are normally a large part of the total. The greatest difficulty with both methods is that of estimating the benefits or alternative costs. From a practical viewpoint both methods give results which are within the probable limits of accuracy.

Table 27.4 Cost allocation for a multiple-purpose project costing $1,765,000

(All items in thousands of dollars)

| Line | Item | Flood mitigation | Power | Irrigation | Navigation | Total |
|---|---|---|---|---|---|---|
| 1 | Separable costs | 380 | 600 | 150 | 50 | 1,180 |
| 2 | Estimated benefits | 500 | 1,500 | 350 | 100 | 2,450 |
| 3 | Alternate single-purpose cost | 400 | 10,000 | 600 | 80 | 2,080 |
| | Remaining benefits method | | | | | |
| 4 | Benefits limited by alternate cost | 400 | 1000 | 350 | 80 | 1,830 |
| 5 | Remaining benefits | 20 | 400 | 200 | 30 | 650 |
| 6 | Allocated joint costs | 18 | 360 | 180 | 27 | 585 |
| 7 | Total allocation: | | | | | |
| | Dollars | 398 | 960 | 330 | 77 | 1,765 |
| | Percent | 22.5 | 54.4 | 18.7 | 4.4 | 100 |
| | Alternative justifiable expenditure method | | | | | |
| 8 | Alternate cost less separable cost | 20 | 400 | 450 | 30 | 900 |
| 9 | Allocated joint costs | 13 | 260 | 292 | 20 | 585 |
| 10 | Total allocation: | | | | | |
| | Dollars | 393 | 860 | 442 | 70 | 1,765 |
| | Percent | 22.2 | 48.7 | 250 | 4.0 | 100 |

In order to establish a pricing policy it is sometimes necessary to allocate costs between several classes of users. When a long pipeline is constructed, it may be proper to charge users near the head of the line less for water than is charged to users at the far and of the line. In this case the allocation is often based on the proportional use of facilities.

No single method of cost allocation can properly be described as the "best" method. For establishing prices or allocating charges to various beneficiaries, any method which is agreeable to all concerned can the considered acceptable.

## New Words and Expressions

1.suffer *v.* 遭受，受损失(from, for, with);
 -er 受害者
2.hypothetic *a.* 假设的
3.arbitrary *a.* 任意的
4.flood-mitigation works 缓解洪水过程，防洪工程
5.benefit-cost ratio 益费比 (or ration of benefit to cost)
6.rate of return 收益率
7.requir of M N 需要 M（干，做出）N
8.in a sense 在某种意义上说
9.cost allocation 费用分摊
10.separable costs 可分费用
11.fish ladders 鱼道
12.joint costs 共同费用
13.remaining-benefits method 剩余效益法
14.alternative-justifiable-expenditure method 合理替代支付法

# Lesson 28 Soil Erosion and Soil Conservation

All civil engineers know two main types of soil movement, soil creep or the slow flow of soil, and the faster flows, called slides, slips or avalanches. The faster movement generally moves layers that are deeper than those moved by soil creep.

But more important than either of these two types (at least in the quantity of soil removed every year) are soil erosion, the removal of soil by flowing water from ground that is usually dry, and scour, the removal of soil or rock from the bed or bank of a river or stream. Forest soil under growing trees generally does not erode because the leaves, and dead or broken wood hold up the rain like blotting paper and prevent its attacking the soil. The main action of dead leaves in delaying runoff is that they hold open the drainage channels into the soil. In bare soil that is not protected in this way, the soil pores are quickly blocked during rain by the fine soil grains flowing into them. The rain runs over the surface and not into it, small channels form, to be deepened eventually into the really damaging deep, scoured ditches called gullies.

The first disadvantage of soil erosion to mankind is that eroded land is so poor that crops will not grow on it, farmers are ruined, and have to move to better land. This has happened to many thousands of farmers in the United States, which in the last hundred years has probably suffered more from erosion than any other country. The second disadvantage is that soil erosion makes flooding both more severe, and more frequent since the water runs off the soil more quickly. In forest areas, 95 percent of the rainfall is absorbed while on bare soil about 95 percent may run off. This proportion is called the runoff coefficient. Nearly all civil engineering structures, roofs, roads, and any concrete work have a runoff coefficient of about 95 percent.

The damage caused by the soil deposited during flooding in a city street has often been calculated, but that caused by the loss of the same valuable soil from the farmland has not, and to the farmer it is equally serious.

With the rapid surface flow caused by the cutting down of the forests, deep channels quickly form in ditches that formerly were small, increasing the speed of erosion in the ditch and forming a gully. The main ways of preventing erosion lie in good farming practice, such as terracing, strip cropping, and ploughing horizontally (contour ploughing) instead of up or down a slope. We cannot study these in this book, but the civil engineer should be aware of erosion and do what he can to help the farmer. Almost all civil engineering work increases the runoff but civil engineers can try to reduce it by keeping their earth slopes gentle and planted with grass or bushes or trees, particularly in upland areas where slopes are steep and erosion can be most serious.

When roads, railways or building foundations break into sloping ground or ground at the foot of a slope, every engineer watches for the danger of earth slips and designs his structures to avoid them but he should also try to conserve soil.

The United States Soil Conservation Service has introduced beavers in forest areas to reduce the flow of rivers during heavy rainfall because these animals build dams across the streams, which hold back water at little or no cost to mankind.

Soil erosion is the cause of nearly one third of the work of maintaining the main roads in the United States. Similar proportions (or higher) probably apply to roads in all hilly country areas where the rainfall is moderate to high. Particular care is needed in the design and building of culverts because the water can scour both the upstream and the downstream side of the culvert. If the culvert slopes steeply downstream, scour can be prevented by building a drop outlet from the culvert (Fig. 28.1). This is an outlet with a waterfall into a concrete structure which the falling water cannot scour. An upstream drop inlet is also possible. Tree planting should be so arranged that it does not cause snow to form deep banks.

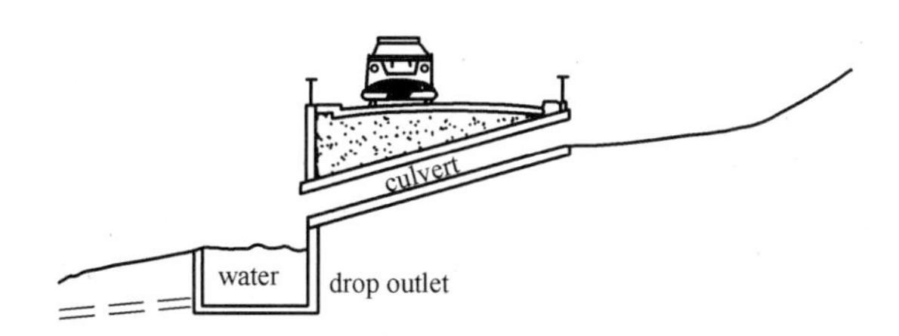


Fig. 28.1 Culvert under a highway on a steep hillside.

The hillside is protected from scour by a drop outlet, a concrete pit with a smooth overflow. It is drained by a small pipe drain if it is necessary to keep down mosquitoes.

The toes of stream banks can be cheaply protected from scour by planting trees which like water. These break the force of the water and encourage the deposition of silt. Banks that are too bare to be planted can be protected by laying matting made of brushwood or rushes over them, in which silt can collect and plant growth can start. For the urgent protection of bare banks exposed to strong currents, either large stones or pre-cast concrete slabs or concrete bag-work may have to be placed.

Jetty protection of banks is slightly different. Short jetties or piers are built out into the river, upstream of the bank being scoured. The disadvantage of jetties is that even if they do prevent scour on their own bank, they may encourage a corresponding scour on the opposite bank. Jetties may be built from piles driven into the river bed, or by tipping large stones into it, or even with rubbish such as old car bodies.

In downstream areas where the river passes through a broad gentle flood plain, civil engineers may be asked to build flood protection works. One such structure is a levee, usually consisting of a long cheap earth bank thrown up as far as conveniently possible from the riverbank. The flood is thus allowed to spread into the land between the river and the levee. The larger the distance between the river and the levee, the greater will be the volume of flood water required to fill it, and the more useful will be the levee. The land thus flooded is usually pasture or other farmland which is not severely damaged by flooding.

## New Words and Expressions

1.slip *v.*，*n.* 滑（动，移，行，过，脱，入，倒），溜（过，走），空转
2.avalanche *n.*，*v.* 上崩，山崩，崩塌，塌方，拥至
3.layer *n.* 层（次），分层，岩层，设计者，压枝，压条，产蛋鸡；*v.* （使）成层，分层，打底
4.remove *v.* 移运，拆卸，除去；*n.* 移动，距离，程度
5.removal *n.* 移动，除去，调动
6.scour *n.* 冲刷(作用)，疏浚，(边，岸)侵蚀，（摩）擦；*v.* 擦，洗，冲刷，疏浚
7.blotting *n.* 吸去（干，墨，油），涂（抹）去
8.attack *n.*，*vt.* 侵袭，腐蚀，投入
9.delay *v.* ，*n.* 抑制，减速，推迟，延迟，滞后
10.pore *n.* 细（微，气管）孔，孔（间）隙，缝；*v.*注视，钻研
11.block *vt.* 堵（阻，闭），塞，断路（流），中断，阻挡（止）
12.gully *n.* 集水沟，排水沟，沟壑，沟渠，雨水进水口，路沟窖井
13.ruin *v.*（使）毁灭，灭亡，残破，(使）变成废墟
14.suffer from 遭受，受到（…之害）；因…而受到损害；(具）有（缺点）的
15.cut *v.*，*n.* 切割，截，剪辑，分割
16.beaver *n.* 水獭（皮），海狸（皮）；*v.*埋头苦干
17.The United States Soil Conservation Service 美国土壤保持总署
18.culvert *n.* 涵洞，暗渠，电缆管道
19.keep down 抑（控）制，消除，减少，缩减
20.mosquitoes *n.* 蚊子 *a.*蚊式的，小型的
21.bagwork *n.* 装袋工作，沙包
22.jetty *n.* （突）码头，突（防波，导流）堤，栈桥，建筑物的突出部分
23.pier *n.* 突（防波）堤，突栈桥，桥台（脚），(凸式）码头，桥墩
24.pile *n.* 桩，(桩）柱，电堆
25.levee *n.*（大，河，天然，冲积，防洪）堤，堤防，码头
26.drive into 敲（打，压，扎，嵌）入
27.throw up 抛（推）上，举起，吐出，丢弃，匆匆建造，产生
28.to hold up 阻滞，拦截
29.to hold open 保持畅通
30.eventually 终于，最后
31.really damaging 真正能起破坏作用的
32.farmers are ruined 农民陷于破产，倾家荡产
33.runoff coefficient 径流系数
34.lie in 在于
35.to be aware of 意识到，认识到，了解
36.do what he can 力所能及地
37.hold back 阻挡
38.with a waterfall 以瀑布方式，像瀑布那样
39.concrete bag work 袋装干拌混凝土
40.as far as conveniently 尽量比较远

## Reading Material Wind Erosion Hazard

Wind erosion is the process by which loose surface material is picked up and transported by the wind, and surface material is abraded by wind-borne particles. The spatial redistribution and resorting of particles by wind erosion may have profound effects on the effected soils, their related micro-topography, and any agricultural activity associated with them.

The process operates in a variety of natural environments that lack a protective cover of vegetation, and it is particularly significant in both not and cold deserts, coastal dune areas, and exposed mountain regions. But its human consequences are undoubtedly most serious in those agricultural areas that experience low, variable and unpredictable rainfall, high temperatures and rates of evaporation and high wind velocity, as is the case in semiarid areas, as well as some of the more humid regions that experience periodic droughts. In such areas the natural process of wind erosion may be accelerated by imprudent agricultural practices, and there may follow a number of physical effects, including soil damage, crop damage and related problems, and numerous undesirable economic consequences.

The record of wind erosion in agricultural areas extends back into classical antiquity, but the hazard achieved international notoriety with the advent of serious soil erosion and related dust and sand storm during the prolonged droughts of the 1930s in the High Plains of North America. As is so commonly the case with human responses to environmental hazards, it was these extreme, spectacular events, accompanied as they were by serious soil disruption and human misery and by much publicity, that provided the mayor impetus for vigorous research into the nature of the wind erosion control methods.

The rate at which wind erosion occurs, if at all, depends on the erodibility and the erosivity of the wind. Wind-tunnel experiments have been concerned mostly with relatively simple situations in terms of erodibility and erosivity, and this has enabled the influence of various factors to be quantitatively assessed.

The erodibility of individual grains is dependent upon their diameter, density, and shape. Most soils, however, consists largely of clods comparing individual particles held together by various forces. It is the state and stability (against abrasion) of these structural units which largely determines the erodibility of soil in a field. If a soil is wellstructured, the number of soil particles small enough to be moved may be very low and abrasion may be minimal due both to a limited supply of abrasives and to the mechanical strength of the structural units. On the other hand, soils with weak structures and ample initial supplies of erodible material may be rapidly abraded. The state and stability of the structural units are principally determined by water, soil texture, organic cements, and disaggregating processes.

The principal factor affecting erosivity is force of the wind on the ground surface. The factors affecting this force can be grouped into two main categories: those relating to the nature of atmospheric flow itself and those relating to the main constraint on that flow, surface roughness.

## New Words and Expressions

1.abrade *vt.* 擦掉，磨损（表皮等），刮除

2.windborne *v.* 风携带；风负荷；风负载

3.resort *v.* 凭借，求助，依赖，诉诸；常去；*n.* 凭借，所凭借的人或物；常去之处

4.notoriety *n.* 臭名昭著，声名狼藉（的）

5.imprudent *a.* 轻率的，不谨慎的；*n.* 轻率的行动

6.antiquity *n.* 古代，古远；古物，古迹

7.advent *n.* 来到，来临

8.impetus *n.* 动力，原动力；刺激，推动力

9.prolong *v.* 延长，拖延

10.spectacular *ad.* 壮观的，引人入胜，洋洋大观的；*n.* 奇观，壮观

11.erosivity *n.* 腐蚀，侵蚀

# Lesson 29 Groundwater and Climate Change

Evidence is mounting that we are in a period of climate change brought about by increasing atmospheric concentrations of greenhouse gases. Global mean temperatures have risen 0.3–0.6 °C since the late 19th century and global sea levels have risen between 10 and 25 cm (Intergovernmental Panel on Climate Change, IPCC, 1995). The IPCC has reported that the expected global rise in temperature over the next century would probably be greater than observed in the last 10,000 years. As a direct consequence of warmer temperatures, the hydrologic cycle will undergo significant impact with accompanying changes in the rates of precipitation and evaporation. Predictions include higher incidences of severe weather events, a higher likelihood of flooding, and more droughts.

In Canada, and particularly in British Columbia (BC), most research on the potential impacts of climate change to the hydrologic cycle has been directed at forecasting the potential impacts to surface water, specifically the links between glacier runoff and river discharge. Relatively little research has been undertaken to determine the sensitivity of aquifers to changes in critical input parameters, such as precipitation and runoff, despite the fact that groundwater constitutes a significant proportion of freshwater supply in Canada. Internationally, only a few studies have been reported in the literature on the impacts of climate change (based on predictive scenarios) to groundwater resources.

The purpose of this study was to assess the sensitivity and to identify potential impacts of climate change (based on projected scenarios) on groundwater in an unconfined aquifer at Grand Forks, BC, Canada. Grand Forks is located in south-central BC, 522 km east of Vancouver, along the Canada–United States border as shown in Fig. 29.1.

The region is semiarid and, because residents rely on groundwater for both domestic and irrigation supply, it is essential that the potential impacts of climate-change on the aquifer be identified for long-term water-management decisions.

Groundwater constitutes approximately 22% of drinking water use in BC and is used for irrigation in many agricultural regions of the province. Therefore, it is important to consider the potential impacts of climate change on groundwater systems. As part of the hydrologic cycle, it can be anticipated that groundwater systems will be affected by changes in recharge (which encompasses changes in precipitation and evapotranspiration), potentially by changes in the nature of the interactions between the groundwater and surface water systems, and changes in use related to irrigation.

In a comprehensive study of groundwater recharge, Vaccaro (1992) used a daily recharge model to estimate the changes in recharge in the semiarid Ellensburg basin on the Columbia plateau in Washington state, USA. He used a stochastic weather generator and a scenario based on the average of three GCMs, and also considered effects under both the pre-development land use (grassland, sagebrush, and forest) and current land use (irrigated and non-irrigated land). Under pre-development land use, recharge would increase by around 10%, both because precipitation increased and because there was increased infiltration from the spring snowpack. With the current land use, recharge is nearly four times higher at present than under pre-development land use largely because of the infiltration of irrigation water. Under the climate-change scenario, recharge with the current land use would reduce by 40%, despite increased rainfall. This is because more of the irrigation water is evaporated and also because there would be less extra infiltration from the snowpack, which is less extensive under the current land use than with the pre-development land cover.

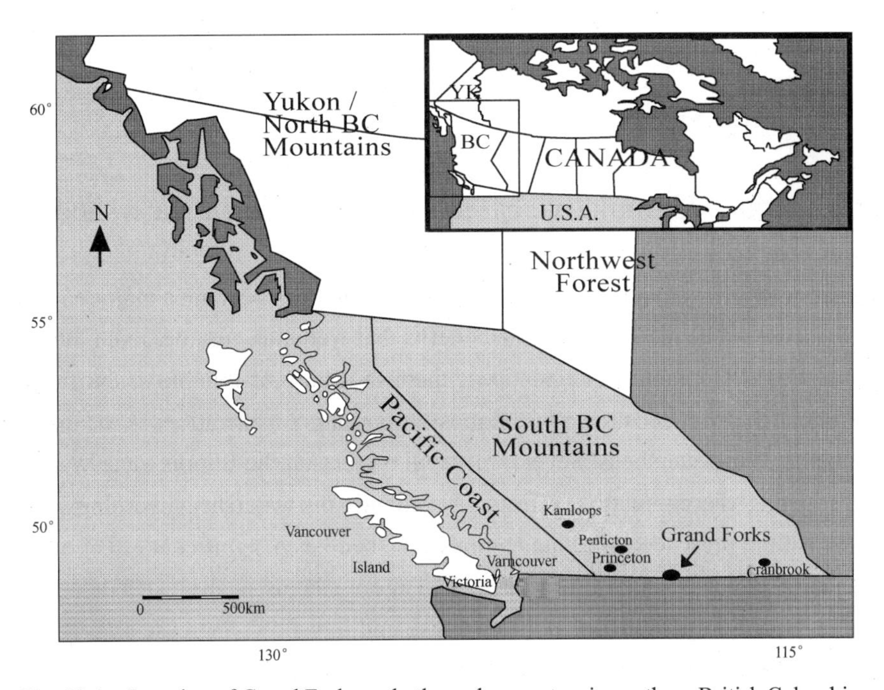


Fig. 29.1 Location of Grand Forks and other urban centers in southern British Columbia.

Recharge to many aquifers in the interior of BC occurs during the late spring and early summer, corresponding to snowmelt. The volume of recharge, therefore, is dependent upon the amount of winter precipitation, the amount of spring snowmelt, and the duration of the recharge season. With no change in the duration of the recharge season, an increase in precipitation would tend to increase recharge, although it is possible that a larger proportion of the extra precipitation would not enter groundwater storage, but would instead run directly off into rivers. The storage of water in the ground is limited by the amount of available pore space; excess water will not infiltrate and will move as surface runoff. Similarly, the infiltration rate is limited according to the type of subsurface material and extreme rainfall events may not result in complete infiltration. A change in recharge to an aquifer would have an effect on groundwater levels, with a consequent effect on river base flow.

In many parts of BC, highly productive aquifers are found in alluvial valleys and consist of unconsolidated deposits of glacial or fluvial origin. Many of these aquifers have rivers across their surfaces. Because of the highly permeable nature of the surficial deposits, strong groundwater–surface water interactions can be anticipated. Therefore, changes to the timing and amplitude of stream/river discharge, which may be associated with climate change, may have considerable impact on groundwater levels.

## New words and Expressions

1.mount *v.* 登上，爬上；增加，证明，组织，安排
2.likelihood *n.* 可能性，似然性，或然性
3.glacier runoff *n.* 冰川径流
4.aquifer *n.* 地下水，地下含水层
5.parameter *n.* 参数，因子
6.stochastic weather generator *n.* 随机气象发生器
7.scenario *n.* 背景，条件，设想的未来事件
8.sagebrush *n.* 山艾，（产于北美西部半荒芜地区）
9.recharge *n.* 补给，地下水补给
10.alluvial valleys *n.* 冲积峡谷，冲积流域
11.unconsolidated *a.* 非固结的，松散的
12.fluvial *a.* 河流的，河川的

## Reading Material Cumulative Effects of Wetlands on watershed hydrology

The conservation of wetlands that have not been drained has become a major environment issue in the world, because of the realization of the benefits that wetlands provide. Even though progress is being made in the development of regulations, there is a lack of reliable scientific procedures to evaluate the function of wetlands. One of the most important considerations is the cumulative effect of dispersed wetlands in watersheds. Several research projects at the Illinois State Water Survey have been on investigating the cumulative effects of wetlands on watershed hydrology. Results of the research show that the presence of wetlands in a watershed has a significant effect on the watershed hydrology, and the effect depends on the extent of the wetlands in the watershed. In general, the research results show that peak flood flows decrease and low flows increase as percent wetlands in the watershed increases. There are, however, significant regional and seasonal differences on how wetlands influence the watershed hydrology.

The conservation of wetlands has become a critical issue because of the rapid rate of wetland losses and the realization of the many benefits wetlands provide. In the mid-1970s it was estimated that more than half of the original wetlands in the United States had been converted to agricultural and urban areas. The total wetland area was estimated to be 99 million acres, representing about 5 percent of the land surface of the United States. The net loss rate of wetlands was estimated to be 458,000 acres per year.

Because of the alarming rate of wetland losses, several regulations and executive orders have been passes by the federal government to protect and preserve wetlands. Several States and local government units have also passed their own regulations dealing with wetlands. In Illinois, legislation passed in 1989 set the goal of “no overall net loss” of wetlands in the state, similar to the federal government’s goals.

Significant progress has been made in developing regulations for wetland protection. However, the implementation of many of the regulations is a hotly contested matter all over the country. The major cause of most of the controversy is the lack of good scientific and engineering information and data. The most important area where the data and information are still inadequate is the field of hydrology and hydraulics. Even though many of the wetland functions related to hydrology and hydraulics are generally understood conceptually, detailed and accurate information and data are not available. The most generally mentioned wetland functions are related to reduction of flood peaks; increase in low flows; ground water recharge; entrapment of sediment, nutrients, and pollutants; and improvement in water quality. However, our knowledge of the relations between the type, size, and location of wetlands and their different hydrologic functions is not very good.

Hydrology is the primary driving force in wetland dynamics. Even though plant species and soil characteristics are generally used as criteria for identifying wetlands, the dominant feature of wetlands is the presence of excess water either on the surface or underground. The existence of any wetland depends on how wet the soil is and/or how long the area is inundated by water. The amount of water available in an area at different times and the way it moves in and out of a wetland area are defined by the hydrology of the area and the hydraulics of flow. When the existing hydrologic regime is altered, the nature and functions of wetlands are altered.

Research is being conducted at the Illinois State Water Survey to better understand the hydrologic functions of wetlands. One of the major components of the research is the quantification of the cumulative effects of wetlands on watershed hydrology. How does the presence or absence of wetlands in a watershed affect the hydrologic characteristics of a watershed? What are the relations between the areal extent of wetlands in a watershed and its hydrologic response? Results from research conducted in Illinois show that the presence of wetlands in a watershed does have significant effect on the watershed hydrology and the significance of the effect depends on how much wetlands are present in the watershed.

There are four major mechanisms by which wetlands can influence streamflows; runoff generation, flow retardation due to increased flow resistance, flood storage, and low-flow augmentation. Runoff generation is linked to wetlands soil moisture and land-cover conditions, important parameters that greatly influence how much runoff is generated from various rainfall events. For example, if an area is already saturated from antecedent rainfall, even small amounts of precipitation can produce large surface runoff. On the other hand, if an area is dry, large amounts of precipitation may produce little or no runoff as most of the precipitation infiltrates into the ground. An area with dense vegetation will intercept more incoming rainwater and have less surface runoff than an area with little or no vegetative cover. Delayed flow from an area with dense vegetation will be greater than from an area with less vegetative cover, however.

The presence of wetland vegetation in the watershed and along drainage channels and streams will increase flow resistance over land and in the stream channels, resulting in slow delivery of runoff from upstream to the watershed outlet. When a flood event occurs along a stream in a watershed with wetlands along the stream banks, the streamflow increases and eventually overflows the banks. The water then has to flow through the adjoining wetlands, which significantly increases the resistance to flow. Since wetlands typically support a large amount of vegetation, the resistance to over bank flow is greater for a stream adjoining a wetland than for a stream not adjoining a wetland. As a result, a significant difference will exist between the flow characteristics of a stream adjacent to a wetland and one not adjoining a wetland.

Another major influence that wetlands exert on streamflow is the flood storage capacity they provide. Since most upland wetlands are located in depressions and poorly drained areas, they provide significant flood storage potential. By storing runoff from adjoining areas, wetlands retard or reduce the amount of runoff that reaches stream channels. Therefore, depending on initial conditions and rainfall amounts, wetland areas could significantly affect the streamflow.

The presence of wetlands in a watershed may also result in low-flow augmentation in streams. In wetland areas, water stored in surficial depressions and underground is gradually released to adjoining streams during periods of low flows. This generally results in higher low-flow conditions for streams that drain areas containing wetlands.

In most watersheds, wetlands are found in several locations dispersed throughout the watershed. Some are located in upland areas while some are located along food plains adjacent to the streams and rivers. The function of each cluster of wetlands might be easy to quantify if detailed information of the topography, soils, and vegetation is available. On the other hand, evaluating the cumulative effects of all the wetlands in the watershed on the watershed hydrology and other functions is very difficult. Policy makers and resource managers often ask the question "how much difference would it make overall if we drain or restore certain areas of wetlands?" This question is very difficult to answer because we do not have an effective procedure for evaluation cumulative impacts.

Conceptually, the presence of wetlands in watershed influences the flood hydrograph in a similar fashion as urban area would but in the opposite direction. As the wetland area in the watershed increases, the peakflow decreases. By how much the peak decreases will depend on several factors such as geology, topography, season, and precipitation intensity and duration. In an attempt to quantify the cumulative effects of wetlands on watershed hydrology, the Illinois State Water Survey has been conducting research in different parts of the state.

## New Words and Expressions

1.conservation *n.* 保护，保守
2.wetland *n.* 湿地
3.cumulative *a.* 累积的，综合的
4.Illinois State Water Survey 美国 Illinois 州水利调查局
5.peak flood flow 洪峰流量
6.hotly *ad.* 激烈地，热烈地，炎热地
7.contested *a.* 竞争的，竞赛的，争夺的，拥挤的
8.controversy *n.* 争议，长期争议
9.nutrient *n.* 营养，养分
10.entrapment *n.* 诱骗，陷阱
11.pollutant *n.* 污染物
12.criteria *n.* 标准
13.dominant *a.* 主要的，占优势的，支配的，统治的
14.inundated *a.* 淹没的
15.mechanism *n.* 结构，过程，方法，机械

16.augmentation *n.* 增加，增长，扩大

17.runoff generation 径流生成，产流

18.antecedent rainfall 前期影响雨量，前期土壤含水量

19.adjoining *a.* 贴近…，与…相临近

20.depressions *n.* 洼地，低洼区，

21.flood hydrograph 洪水过程线

22.peak flow = peak flood flow *n.* 洪峰流量

23.precipitation intensity 雨强，降雨强度

# Lesson 30 Irrigation Performance Evaluation

Assessment of irrigation performance is a prerequisite for improving water use in the agricultural sector to respond to perceived water scarcity. Between 1996 and 2000, we conducted a comprehensive assessment of the performance of the Genil–Cabra Irrigation Scheme (GCIS) located in Andalusia, southern Spain. The area has about 7,000 ha of irrigated lands distributed in 843 parcels and devoted to a diverse crop mix, with cereals, sunflower, cotton, garlic and olive trees as principal crops. Irrigation is on demand from a pressurized system and hand-moved sprinkler irrigation is the most popular application method. Six performance indicators were used to assess the physical and economic performance of irrigation water use and management in the GCIS, using parcel water-use records and a simulation model. The model simulates the water-balance processes on every field and computes an optimal irrigation schedule, which is then checked against actual schedules. Among the performance indicators, the average irrigation water supply:demand ratio (the ratio of measured irrigation supply to the simulated optimum demand) varied among years from 0.45 to 0.64, indicating that the area is under deficit irrigation. When rainfall was included, the supply:demand ratio increased up to 0.87 in one year, although it was only 0.72 in the driest year, showing that farmers did not fully compensate for the low rainfall with sufficient irrigation water.

The availability of water for irrigation will probably decrease in the future due to increased demands from other sectors, such as municipal, tourism, recreation and the environment. In Spain, fresh-water demand is estimated as $35\times10^5$ $m^3$/year with about 70% devoted to irrigation and the rest to other uses. Additionally, the government anticipates that irrigation demand in Southern Spain will increase by about 17% in the next 10 years.

Improvements in water management and the modernization/rehabilitation of the Spanish irrigation schemes are important objectives to achieve more efficient use of water. Only 27% of the irrigated area in Spain (approximately 915,000 ha) is less than 20 years old, whereas 37% is more than 90 years old. In recent years, the water administration emphasized system modernization and rehabilitation but comparatively little attention was paid to the improvement of irrigation management.

The improvement of water management in an irrigation scheme requires the assessment of irrigation performance as a point of departure. Computer simulation using hydrologic models has been useful for this task. Many models have been used to simulate parts of the hydrologic cycle in irrigated agriculture, from empirical or functional to mechanistic. Additionally, to facilitate data acquisition and carry out spatial analyses, recently developed tools, such as remote sensing and geographic information systems, have been combined with hydrologic models to assess the behavior of irrigation schemes.

Several authors have defined sets of indicators that characterize irrigation system performance, intending to evaluate current practices and recommend improvements in irrigation efficiency and water productivity. These performance indicators are also used to quantify the system ability and to achieve the objectives established for an irrigation area or to assess the current performance of the system relative to its potential.

The different types of performance indicators are related to: ①the water balance, ②economic, environmental and social objectives, ③system maintenance. Several authors have used these indicators for: ①assessing trends in performance, ②comparing performance among irrigation schemes, ③resource optimization and ④ determining a compromise solution between equity and efficiency within an irrigation area. The complexity of models used for calculating the water balance-based performance indicators varies from one-dimensional, physically based, hydrologic model to very simplified models, such as those based on the FAO methodology. At the scheme level, input information to compute performance indicators is normally obtained from total water delivery records and from water consumption estimates derived from the cropped areas. Such information often has substantial uncertainty and does not allow for in-depth analysis at levels below the scheme. Nevertheless, scheme-level assessments are needed for comparative purposes and are the only approach feasible when there is no access to information at sub-scheme levels.

The objective of this work was to conduct a comprehensive assessment of the irrigation performance of an area using on-farm water-use information and a simulation model. The area selected was the Genil–Cabra irrigation scheme (GCIS) located in Andalusia, southern Spain. This area was chosen because it was possible to obtain accurate information on water use and on the cropping patterns of individual parcels during four irrigation seasons.

## New Words and Expressions

1.assessment *n.* 评价，评估
2.prerequisite *n* 先决条件
3.water scarcity *n.* 缺水
4.cereals *n.* 谷物，粮食作物
5.parcels *n.* 单元，地块
6.olive *n.* 橄榄树
7.pressurized system *n.* 有压系统
8.hand-moved *a.* 手动的，靠手操作的
9.indicators *n.* 指标，标准
10.recreation *n.* 娱乐，疗养
11.anticipate *v.* 预测，预计
12.municipal *a.* 城市的，市区的
13.rehabilitation *n.* 改造，恢复，更新
14.empirical *a.* 经验性的，凭经验的
15.mechanistic *a.* 机械的，结构的，方法性的
16.water consumption *n.* 用水，水量消耗
17.accurate *a.* 精确的，确切的

## Reading Material Economic and agronomic strategies to achieve sustainable irrigation

The achievement of sustainable irrigation in arid regions requires greater attention to waterlogging, salinization, and degradation of ground and surface waters, which are among the problems that continue to threaten productivity and degrade environmental quality. We consider sustainability to be achieved when irrigation and drainage are conducted on-farm, and within irrigation districts, in a manner that does not degrade the quality of land, water, and other natural resources, either on-farm or throughout an irrigated region. Sustainability may also be described as maintaining the productive resources required for irrigation, so that future generations may have the same opportunity to use those resources as we do. Given the increasing importance of irrigated land for food production, the time has come when it is vital to intercept, reuse, and isolate drainage waters within the regions in which they are generated. Adoption of this strategy can be enhanced by policies that require farmers, and irrigation districts, to consider the off-farm impacts of irrigation and drainage. Such policies include linking water rights with salt rights to require the monitoring and management of both irrigation water and the salt loads in drainage waters. We review the knowledge gained since the early 1970s regarding the economic and agronomic aspects of irrigation and drainage, with a focus on drainage water reduction and sequential reuse of drainage water on salt-tolerant crops. Economic incentives that motivate farm-level and district-level improvements in water management are also reviewed. We conclude that adequate knowledge exists for implementing strategies that focus on water use and salt disposal within irrigated regions, and we recommend policies that will motivate improvements in productivity and enhance the likelihood of achieving sustainability.

Only about 17% of the world's cropland is irrigated, but that land produces more than one-third of the food and fiber harvested worldwide. Expansion of irrigation in the period between the mid-1960s and mid-1980s accounted for more than 50% of the increase in global food production during that time. In India, this figure approaches 100%. Expansion of irrigation will need to keep pace with the increasing world population. However, expansion will not be as easy as it was in the past, given the loss of irrigated lands to salinization, and increasing competition for limited supplies of good-quality irrigation water and the rising costs of developing those supplies.

Future water demands can be met, in part, by using available water supplies more efficiently. Increased use of municipal wastewaters and irrigation drainage waters will also become necessary. The salinity (total salt content) and sodicity (sodium content) of these waters will be higher than that of the original source water because of the direct addition of salts to the water and the evapoconcentration that occurs as water is reused. Higher salinities and sodicities impair crop productivity and farm profits, thereby affecting the ability of farmers to remain in business and increasing the inevitable negative off-site environmental impacts of irrigated agriculture .

Increasing the production of food and fiber to support a larger world population requires that we improve the performance and enhance the sustainability of irrigation systems in many regions. Irrigation performance is described by the value of outputs and amenities generated in comparison with the level of inputs and opportunity costs required to operate and maintain an irrigation system. Performance is evaluated by analyzing water deliveries, crop yields, and the market prices of inputs and outputs. Sustainability describes the likelihood that an irrigation system will continue to generate desirable outputs and amenities at reasonable costs in future. Hence, the analysis of sustainability requires a broader framework than is used to evaluate irrigation performance. Sustainability is achieved when irrigation and drainage are conducted in a manner that does not degrade the quality of land, water, and other natural resources that contribute to agricultural production and environmental quality. This view of sustainability may also be described as maintaining the productive resources required for irrigation, so that future generations may have the same opportunity to use those resources as that afforded to the present generation.

We believe that sustainable irrigation can be achieved if issues regarding efficient irrigation, drainage management, and salt disposal are addressed directly by farmers and public officials. The wealth of knowledge regarding the physical aspects of irrigation and drainage in arid regions can be combined with experience regarding economic incentives to design policies that will motivate farmers and irrigation districts to seek long-term, sustainable solutions to the age-old problems of waterlogging and salinization. The solutions should involve farm-level and regional efforts to intercept, isolate, and dispose of saline drainage water, rather than allowing excess water and salts to reduce productivity on lower lying irrigated lands or to degrade the quality of rivers and groundwaters. Intraregional salt disposal on the land surface will require careful management, including mitigation, to minimize agricultural and environmental impacts. Environmental regulations and economic incentives will enhance the rate at which these strategies (intercepting, isolating, and disposing of drainage waters) are implemented. The benefits for users of groundwater and surface waters—down-slope, down-canal and down-river—will include greater agricultural production, improved water quality for municipalities, and enhanced amenity values (i.e. boating, swimming, bird watching, sightseeing, etc.).

We agree with the perspective that serious efforts should be made to motivate the efficient use of land and water resources, particularly in arid regions where waterlogging and salinization and increasing water demands are serious issues. The off-farm impacts of irrigation and drainage involve all three factors described by Professor Biswas, and policies designed to reduce off-farm impacts should address those factors explicitly. Waterlogging and salinization are classic examples of externalities that arise when farmers, or their irrigation districts, are not required to consider the off-farm or downstream impacts of excessive irrigation or the salt loads in surface runoff and deep percolation. Policies that assign responsibility to farmers and irrigation districts for both the source water and salt loads in surface and subsurface return flows from irrigation would likely reduce the areal extent of waterlogging and salinization in arid regions. The probability of success is enhanced when such policies are implemented by a local or regional agency that works with farmers when designing policy measures and selecting parameter values. Examples include the Catchment Management Authorities that address waterlogging and salinization problems in Australia and the Grassland Area Drainage Authority formed to reduce the discharge of agricultural drainage water into California's San Joaquin River.

## New Words and Expressions

1.sustainable irrigation 可持续灌溉

2.waterlogging *n.* 水涝，水涝灾害

3.salinization *n.* 盐碱化

4.degradation *n.* 降低质量，降级

5.adoption *n.* 采用，采取，实现

6.salt loads *n.* 含盐量，盐分负荷

7.salt-tolerant crops 耐盐作物

8.economic incentives 经济措施，经济机制，经济刺激

9.accounted for *v.* 解释，说明，占…（有一定的数量）

10.to keep pace with 与…齐步前进

11.impair *v.* 损坏，破坏，使失去平衡

12.in comparison with 与…相比较

13.amenity *n.* 和谐，舒适，宜人

14.age-old problems 旧问题，老问题

15.intraregional *a.* 区域内

16.perspective *n.* 想法，观点，景观

# Lesson 31 Bidding, Bid Opening and Award of Contract

The bidding documents should state clearly whether contracts will be awarded on the basis of unit prices (for work performed or goods supplied) or of a lump sum of the contract, according to the nature of goods or works to be provided.

The size and scope of individual contracts will depend on the magnitude, nature, and location of the project. For projects requiring a variety of works and equipment such as power, water supply, or industrial projects, separate contracts are normally awarded for the civil works, and for the supply and erection of different major items of plant and equipment.

On the other hand, for a project requiring similar but separate civil works or items of equipment, bids should be invited under alternative contract options that would attract the interest of both smaller and larger firms. Contractors or manufacturers, small and large, should be allowed to bid for individual contracts or for a group of similar contracts at their option, and all bids and combinations of bids should be opened and evaluated simultaneously so as to determine the bid or combination of bids offering the most advantageous solution for the borrower.

Detailed engineering of the works or goods to be provided, including the preparation of technical specifications and other bidding documents, should precede the invitation to bid for the contract. However, in the case of turnkey contracts or contracts for large complex industrial projects, it may be undesirable to prepare technical specifications in advance. In such a case, it will be necessary to use a two-step procedure inviting unpriced technical bids subject to technical clarification and adjustments, followed by the submission of priced proposals.

The time allowed for preparation of bids should depend on the magnitude and complexity of the contract. Generally, not less than 45 days from the date of invitation to bid should be allowed for international bidding. Where large civil works are involved, generally, not less than 90 days from the date of invitation should be allowed to enable prospective bidders to conduct investigations at the site before submitting their bids. The time allowed, however, should be governed by the particular circumstances of the project.

The date, hour, and place for latest delivery of bids by the bidder, and of the bid opening, should be announced in the invitation to bid, and all bids should be opened at the stipulated time. Bids delivered after the time stipulated should be returned unopened unless the delay was not due to any fault of the bidder and its late acceptance would not give him any advantage over other bids. Bids should normally be opened in public. The name of the bidder and total amount of each bid, and of any alternative bids if they have been requested or permitted, should, when opened, be read aloud and recorded.

Extension of validity of bids should normally not be requested; if, in exceptional circumstances, an extension is required, it should be requested of all bidders before the expiration date. Bidders should have the right to refuse to grant such an extension without forfeiting their bid bond, but those who are willing to extend the validity of their bid should be neither required nor permitted to modify their bids.

It is undesirable that information relating to the examination, clarification, and evaluation of bids and recommendations concerning awards be communicated after the public opening of bids to bidders or to persons not officially concerned with these procedures until the award of a contract to the successful bidder is announced.

No bidder should be permitted to alter his bid after bid has been opened. Only clarifications not changing the substance of the bid may be accepted. The borrower may ask any bidder for a clarification of his bid but should not ask any bidder to change the substance or price of his bid.

Following the opening, it should be ascertained whether material errors in computation have been made in the bids, whether the bids are substantially responsive to the bidding documents, whether the required sureties have been provided, whether documents have been properly signed, and whether the bids are otherwise generally in order. If a bid is not substantially responsive to the bidding documents, or contains inadmissible reservations, it should be rejected, unless it is an alternative bid permitted, or requested, under the bidding documents. A technical analysis should then be made to evaluate each responsive bid and to enable bids to be compared.

A detailed report on the evaluation and comparison of bids setting forth the specific reasons on which the decision for the award of the contract, or rejection of all bids, is based should be prepared by the borrower or by its consultants.

The award of a contract should be made, within the period specified for the validity of bids, to the bidder whose responsive bid has been determined to be the lowest evaluated bid, and who meets the appropriate standards of capability and financial resources. Generally, such bidder should not be required, as a condition of award, to undertake responsibilities or work not stipulated in the specifications or to modify his bid.

## New Words and Expressions

1.lump *n.* 堆，团，大量，一大堆；*v.*总括，概括，集中

2.lump sum 总数，总额，总金额

3.civil works 土建，土木工程

4.erection *n.* 建设，安装，装配，竖立

5.turnkey *n.* 包到底的工程，交钥匙工程

6.turnkey contract 包括规划，设计和管理的施工合同，整套承包合同

7.precede *v.* 领先，居前，优先

8.undesirable *a.* 不适当的，不合乎需要的，不方便的；*n.* 不受欢迎的人

9.proposal *n.* 申请，建议，计划，投标

10.stipulate *v.* 规定，限定，做为条件来要求，坚持

11.forfeit *n.* 罚款，没收物；*a.*被没收的； *v.* 被没收，丧失

12.bidder *n.* 出价人，投标人

13.surety *n.* 确实，保证，保证金

14.(be) in order （是）完好的，有条理的，适用的，处于可使用状态

15.inadmissible *a.* 不能允许的，不能承认的，不能采纳的

# Reading Material Competitive Bidding

Competitive bidding based on tender documents prepared by the client's professional advisors is still the most common method of distributing the construction industry's contracts among the contractors willing to undertake the work. Variations such as negotiated contracts and package deals form only a proportion of the contracts offered to the industry. The acceptance by the majority of clients, mainly central and local governments, that competitive bidding is fair and will produce the lowest possible commercially viable tender price in the prevailing market conditions, ensures that this form of work distribution will continue for a long time. The random nature of bidding process also ensures that contracting companies will be unable to plan the company's activities with much certainty; that many contracts will be tendered for with unrealistically low prices and that the preoccupancy of most contractor with claims will also continue.

From the contractor's viewpoint competitive bidding has the appearance of roulette: sometimes he wins when he thinks his price is high; sometimes he loses when his price is dangerously low, and he has a wry smile for the apparent "winner". Often when a contractor obtains a contract he resorts to claims to ensure that the achieved mark-up is positive because the original tender was based on a low cost estimate. It is not surprising therefore that the subject of "competitive bidding" has attracted investigations and research by both the contracting companies themselves and a variety of academics.

The subject of bidding strategy has interested various researchers in America and Europe since the mid-1950s. The aim of most of these workers has been the development of a probabilistic model, which will predict the chances of winning in the type of competitive bidding that is common in the construction industry. These probabilistic models have attempted to give guidance to bidders by producing statements of the type.

In entering a bidding competition it is assumed that the contractor first estimates his costs and then adds a mark-up to cover profit (or a mark-up to cover contribution, i. e. profit and company overheads). If the contractor is really desperate to win he could submit a bid at something less than cost. If this bid was low enough then it would have a 100% chance of winning. Just as at the lower end there exists this bid with 100% chance of winning there also exists at the other extreme a bid with no chance of winning (say cost plus 50% mark-up). Between these two extremes there exists a continuum of bids with associated probabilities which measure the chance of winning.

There is obviously a relationship between mark-up and success rate. Increases in the mark-up will reduce success rate. Sometimes the advice given is to increase the applied mark-up and compensate for the reduced success rate by bidding for more contracts. This advice may have a firm theoretical basis but can only be applied if there are the extra contracts available and should only be applied if the sensitivity of success rate to changes in mark-up is known. Changes in mark-up policy of different companies would clearly lead to different outcomes. The outcome for a particular company should be examined before any unconsidered action is taken.

## New Words and Expressions

1.competitive *a.* 竞争的，比赛的
2.tender *n.* 招标，投标，承包，标书；*v.*提出，投标，报价
3.client *n.* 委托人，当事人，买主
4.contract *n.* 合同，契约，承包
5.contractor *n.* 承包人，承包商，立契约人
6.package deal 整批交易，整套工程
7.viable *a.* 能生存的，有生存力的，可行的
8.prevailing *a.* 流行的，主要的，占优势的
9.random *a.* 随便的，偶然的，随机的
10.preoccupancy *n.* 先占，先取
11.appearance of roulette 轮盘下赌的样子
12.wry *v.*扭曲；*a.* 扭歪的，面部肌肉扭曲的，曲解的
13.wry smile 苦笑
14.apparent *a.* 明白的，显然的
15.resort *v.*，*n.* 求助，依靠，凭借，求助的对象
16.mark-up 抬高价格，提高标价，标高金额，加价
17.purport *v.* 意味着，表明，说明
18.strategy *n.* 战略，策略，对策
19.budgeted turnover 预算营业额
20.overhead *n.* 经常费，管理费
21.outcome *n.* 结果，成果，产量，输出

# Lesson 32 How to Write a Scientific Paper

## 1 Title

In preparing a title for a paper, the author would do well to remember one salient fact: That title will be read by thousands of people. Perhaps few people, if any, will read the entire paper, but many people will read the title, either in the original journal or in one of the secondary (abstracting and indexing) services. Therefore, all words in the title should be chosen with great care, and their association with one another must be carefully managed.

The title of a paper is a label. It is not a sentence. Because it is not a sentence, with the usual subject, verb, object arrangement, it is really simpler than a sentence (or at least, usually shorter), but the order of the words becomes even more important.

The meaning and order of the words in the title are of importance to the potential reader who sees the title in the journal table of contents. But these considerations are equally important to all potential users of the literature, including those (probably a majority) who become aware of the paper via secondary sources. Thus, the title should be useful as a label accompanying the paper itself, and it also should be in a form suitable for the machine-indexing systems used by Chemical Abstracts. The Engineering Index, Science Citation Index, and others. Most of the indexing and abstracting services are geared to "key word" systems. Therefore, it is fundamentally important that the author provide the right "keys" to the paper when labeling it. That is, the terms in the title should be limited to those words that highlight the significant content of the paper in terms that are both understandable and retrievable.

## 2 Abstract

An Abstract should be viewed as a mini-version of the paper. The Abstract should provide a brief summary of each of the main sections of the paper. A well-prepared abstract enables readers to identify the basic content of a document quickly and accurately, to determine its relevance to their interests, and thus to decide whether they need to read the document in its entirety. The Abstract should not exceed 250 words and should be designed to define clearly what is dealt with in the paper. Many people will read the Abstract, either in the original journal or in The Engineering Index, Science Citation Index, or one of the other secondary publications.

The Abstract should ①state the principal objectives and scope of the investigation, ②describe the methodology employed, ③summarize the results, ④state the principal conclusions. The importance of the conclusions is indicated by the fact that they are often given three times: once in the Abstract, again in the Introduction, and again (in more detail probably) in the Discussion.

The Abstract should never give any information or conclusion that is not stated in the paper. References to the literature must not be cited in the Abstract (except in rare instances, such as modification of a previously published method).

## 3 Introduction

Now that we have the preliminaries out of the way, we come to the paper itself. I should mention that some experienced writers prepare their title and Abstract after the paper is written, even though by placement these elements come first. You should, however, have in mind (if not on paper) a provisional title and an outline of the paper that you propose to write. You should also consider the level of the audience you are writing for, so that you will have a basis for determining which terms and procedures need definition and which do not.

The first section of the text proper should, of course, be the Introduction. The purpose of the Introduction should be to supply sufficient background information to allow the reader to understand and evaluate the results of the present study without needing to refer to previous publications on the topic. The introduction should also provide the rationale for the present study. Above all, you should state briefly and clearly your purpose in writing the paper. Choose references carefully to provide the most important background information.

Suggested rules for a good Introduction are as follows: ①It should present first, with all possible clarity, the nature and scope of the problem investigated. ②It should review the pertinent literature to orient the reader. ③ It should state the method of the investigation. If deemed necessary, the reasons for the choice of a particular method should be stated.④ It should state the principal results of the investigation. ⑤It should state the principal conclusion suggested by the results. Do not keep the reader in suspense; let the reader follow the development of the evidence.

## 4 Materials and Methods

In the first section of the paper, the Introduction, you stated the methodology employed in the study. If necessary, you also defended the reasons for your choice of a particular method over competing methods.

Now, in Materials and Methods, you must give the full details. The main purpose of the Materials and Methods section is to describe the experimental design and then provide enough detail that a competent worker can repeat the experiments. Many (probably most) readers of your paper will skip this section, because they already know (from the Introduction) the general methods you used and they probably have no interest in the experimental detail. However, careful writing of this section is critically important because the cornerstone of the scientific method requires that your results, to be of scientific merit, must be reproducible; and, for the results to be adjudged reproducible, you must provide the basis for repetition of the experiments by others. That experiments are unlikely to be reproduced is beside the point; the potential for producing the same or similar results must exist, or your paper does not represent good science.

When your paper is subjected to peer review, a good reviewer will read the Materials and Methods carefully. If there is serious doubt that your experiments could be repeated, the reviewer will recommend rejection of your manuscript no matter how awe-inspiring your results.

In describing the methods of the investigations, you should give sufficient details so that a competent worker could repeat the experiments, as stated above. If your method is new (unpublished) you must provide all of the needed detail. However, if a method has been previously published in a standard journal, only the literature reference should be given.

## 5 Results

So now we come to the core of the paper, the data. This part of the paper is called the Results section.

There are usually two ingredients of the "Results" section. First, you should give some kind of overall description of the experiments, providing the "big picture, " without, however, repeating the experimental details previously provided in "Materials and Methods". Second, you should present the data.

Of course, it isn't quite that easy. How do you present the data? A simple transfer of data from laboratory notebook to manuscript will hardly do. Most important, in the manuscript you should present representative data rather than endlessly repetitive data.

The "Results" need to be clearly and simply stated, because it is the Results that comprise the new knowledge that you are contributing to the world. The earlier parts of the paper ("Introduction", "Materials and Methods") are designed to tell why and how you got the "Results"; the later part of the paper ("Discussion") is de signed to tell what they mean. Obviously, therefore, the whole paper must stand or fall on the basis of the Results. Thus, the "Results" must be presented with crystal clarity.

## 6 Discussion

The "Discussion" is harder to define than the other sections. Thus, it is usually the hardest section to write. And, whether you know it or not, many papers are rejected by journal editors because of a faulty "Discussion", even though the data of the paper might be both valid and interesting. Even more likely, the true meaning of the data may be completely obscured by the interpretation presented in the "Discussion", again resulting in rejection.

What are the essential features of a good "Discussion"? I believe the main components will be provided if the following injunctions are heeded:

(1) Try to present the principles, relationships, and generalizations shown by the "Results". And bear in mind, in a good "Discussion", you discuss—you do not recapitulate the "Results".

(2) Point out any exceptions or any lack of correlation and define unsettled points. Never take the high-risk alternative of trying to cover up or fudge data that do not quite fit.

(3) Show how your results and interpretations agree (or contrast) with previously published work.

(4) Don't be shy; discuss the theoretical implications of your work, as well as any possible practical applications.

(5) State your conclusions as clearly as possible.

(6) Summarize your evidence for each conclusion.

In showing the relationships among observed facts, you do not need to reach cosmic conclusions. Seldom will you be able to illuminate the whole truth; more often, the best you can do is shine a spotlight on one area of the truth. Your one area of truth can be buttressed by your data; if you extrapolate to a bigger picture than that shown by your data, you may appear foolish to the point that even your data-supported conclusions are cast into doubt.

When you describe the meaning of your little bit of truth, do it simply. The simplest statements evoke the most wisdom; verbose language and fancy technical words are used to convey shallow thought.

## New Words and Expressions

1.salient *a.* 突出的，显著的，卓越的，优质的，明显的
2.table of contents 目录
3.highlight *n.* 重点，要点；*v.*着重，强调
4.relevance *n.* 关联，关系，适用，中肯
5.preliminary *a.* 初步的，序言的；*n.(pl.)* 正文前面的内容，准备工作
6.provisional *a.* 暂定的，假定的，暂时的，临时的
7.rationale *n.* 基本原理，理论基础，原理的阐述
8.above all 尤其是，最重要的是，首先是
9.orient *v.* 定向，取向，正确地判断，（使）适应
10.obscure *a.* 模糊的，不清楚的，难解的
11.injunction *n.* 命令，指令
12.heed *vt.*，*n.* 注意，留心
13.recapitulate *v.* 扼要重述，概括，重现，再演
14.unsettled *a.* 不稳定的，不安定的，未解决的，混乱的
15.correlation *n.* 关联，相关性，相互关系
16.cover up 包裹，隐藏，掩盖
17.fudge *n.* 捏造，空话；*v.* 粗制乱造，捏造，推诿
18.implication *n.* 纠缠，隐含，意义，本质
19.cosmic *a.* 宇宙的，全世界的，广大无边的
20.illuminate *v.* 照亮，阐明，使明白，使显扬，使光辉灿烂
21.buttress *n.* 支持物，支柱；*v.*支持，加强
22.extrapolate *n.* 推断，外推，外插
23.evoke *vt.* 唤起，引起，博得，移送
24.fancy *n.*，*a.* 想象，美妙的，漂亮的

## Reading Material Technical Report Elements

### 1 Abstract

The abstract is a condensed statement of the important information contained in the complete report. It is the epitome of a summary and is written in the present tense. It stresses the objective and conclusions. An abstract allows the reader to survey the purpose, content, and conclusions of a report quickly. The two most important requirements of an abstract are that it be concise and informative. To accomplish this, the abstract is usually written last. Notes: In some engineering reports, a more extensive summary may be presented instead of an abstract. The summary includes all of the elements of the abstract, plus a summary of the important results.

### 2 Introduction

The primary function of an introduction is to let the reader know the importance of the work and to clearly define the objective. One this is stated, a brief plan of development should follow. A well-constructed introduction should stimulate reader interest and summarize the contents of the report. Background information of a theoretical or historical nature may be warranted to support this preliminary information. As you would expect, the introduction is that section that introduces the work to the reader. The beginning of the introduction usually explains the problem and the objective of the report. Journalists are taught to answer the five W's: who, what, where, when and why. It is probably a good idea to keep these in mind when writing your introduction and to answer those W's that are pertinent to your objective. For example, the WHO might be the names of previous investigators that you found in a literature search. The WHAT would be the problem statement. The WHY might be pertinent applications of your results. The WHERE and WHEN are obvious.

The introduction, like the rest of the report, is basically written in the past tense. The use of the present tense is limited to material internal to the report.

The introduction is used to acquaint the reader with the material of the report. As part of this presentation it is advisable to state some of the important principles of the work and enumerate assumptions.

## 3 Analysis

The analysis section is used to develop a pertinent theory based on the basic principles that explain the phenomenon you are investigating. Most experimental studies involve the interaction of a variety of complex influences and subtleties. The purpose of the analysis is to remove the mask of complexity and expose the underlying facts. It is a process of systematic thinking, combining logical assumptions with basic principles to develop a relationship that explains your results. This relationship is usually the hypothesis that is the subject of the report. The experiment is the study of this hypothesis to test if your assumptions and logic correct.

The analysis is a written text usually interspersed with equations. It is not simply a series of equations devoid of explanatory material. The explanation of technical material is naturally associated with mathematics. Assumptions, which are expressed in words, are transformed into their mathematical equivalents. Basic principles are also expressed in mathematical terms and are combined with the assumptions to develop the hypothesis. Intermediate steps showing the algebra and calculus, while necessary to the development of the hypothesis, are not shown. However, your presentation should be complete enough that a peer could duplicate your work. It is frustrating to see the expression, "it can readily be seen that" between two equations which bear no apparent relationship one to the other. If some real detail is necessary to fully explain a particular point but is extraneous to your basic presentation, then this work belongs in an appendix.

Equations must be presented clearly with explanatory material relating the equation to the remainder of the report. Symbols should be defined when they are first introduced. All the symbols in an equation must be defined. However, it is not necessary to redefine terms once they have been presented. If the report contains a number of unfamiliar symbols, give a nomenclature section. This section lists the symbols in alphabetical order along with their description in addition to the usual definitions in the text when the terms are first presented.

## 4 Procedure

This section describes the apparatus and details the experimental procedure for taking measurements. In this section, you must explain what was measured and how you measure it. You should provide sufficient detail so that the experiment can be replicated using the same or equivalent equipment. Drawings showing the setup are often useful. They can be an aid in describing certain measurements and they should show the interconnections of the various instruments.

The procedure does not contain results. You can explain that 20 separate tests were performed. You can say that the means and standard deviation were calculated, but you do not give the numerical values. These values are presented in the Results Section.

## 5 Discussion of Results

Results are the facts. They are the data you collected and the data you calculated. Means, standard deviations, confidence intervals and errors are all results.

Present the results in a logical and concise fashion. You can place sample calculations in this section. But if you want detail and an explanation of a series of extraneous calculations, then use an appendix. In general, the detailed calculations of the experimental errors are best placed in an appendix unless the analysis of the errors is the object of the report.

Do not transcribe your raw data. These are the numbers you recorded from your experimentation. A xerox or carbon copy of this data should be in the appendix forming the last page of your report. Thus, it is important to keep a neat, clear and informative laboratory notebook, and all your lab partners could have the same identical last page of their report.

In the procedure, you explained how and what were measured. Now you give the results. The results are the facts; given the same raw data, the reader should get the same results. Repeating the experiment should give similar results. But even when the results are identical, readers may interpret the results differently weighing certain information more heavily. These interpretations of the results are called conclusions.

## 6 Conclusions

It is interesting that, given the same results, two people can draw two different conclusions, and neither conclusion is necessarily incorrect. That is not to say that any conclusion is correct but that a conclusion is personal; it is your interpretation of the results and is subjective. However, the conclusion should relate to the objective of the report.

Students hesitate to make conclusions for fear of being wrong. "This method of determining the coefficient of friction was a reasonably good way of obtaining fairly accurate results," says nothing. It straddles the issue and avoids being wrong. It is better to be decisive when the results warrant a decision.

Some legitimate conclusions are:

(1) This experiment showed that the coefficient of static friction between aluminum and brass is not a simple value but can vary by as much as 50%.

(2) This experiment showed that it is not necessary to use sophisticated or expensive equipment to obtain accurate results.

(3) For an experiment stressing precision, the equipment was unusually crude. No wonder the results had such variability. Better equipment would have given more precise answers.

All these conclusions may be valid; it depends upon the results. Remember that conclusions are not facts. They are your interpretation of the facts, and these interpretations should pertain to the objective of the report. They should bring your report to a sensible finish.

# New Words and Expressions

1.condense *v.* 浓缩，压缩，简要叙述
2.epitome *n.* 梗概，概括，缩影，集中表现
3.enumerate *v.* 数，枚数，列举，计算
4.subtlety *n.* 稀薄，精巧，微妙，微细
5.hypothesis (*pl.* hypotheses) *n.* 假说，假设，前提
6.intersperse *vt.* 散布，散置
7.extraneous *a.* 外部的，附加的，无关的，不重要的
8.nomenclature *n.* 名词，术语，术语表
9.confidence interval 置信区间
10.transcribe *vt.* 抄写，记录，改编
11.straddle *v.*，*n.* 对…不表态，骑墙，观望
12.legitimate *a.* 合法的，正规的，合理的，真实的
13.Xerox or carbon copy 复印或复写本，副本
14.sensible *a.* 切合实际的，合理的，可感觉的
15.acquaint *v.* 使熟悉，使知道，通知
16.decisive *a.* 确定的，决定的，果断的，明确的
17.replicate *vt.* 重复，复现，重制； *a.* 重复的，重复的实验

# 中文论文的英文摘要写作要求与方法

摘要是科技文献内容有关要点的概述，其目的是对一个报告或者任何出版物、发表的论文等作出简明扼要的概括总结，不需进入细节，用于帮助调查研究人员对大量的文献资料尽快而有效地找出有关的信息。

联合国教科文组织规定：“全世界公开发表的科技论文，无论用哪种文字写成，都必须附有一篇短小的英文摘要。”因此，我国现有的“公开发行”或部分“限国内发行”的学术刊物，都要有中文摘要，而且应附有相应的英文摘要，其目的是为了扩大对外交流，逐步实现与国际学术交流接轨，并使高水平的论文尽可能多地被国外检索机构收录。

英文摘要一般要求附在中文摘要之后，也有的刊物要求附在中文论文的最后，它包括：论文的标题、作者姓名（汉语拼音）及单位（要注出邮政编码）、摘要和关键词。

## 1　英文标题（Title）

科技论文的标题是论文内容的高度概括，应以最精炼的文字，充分反映论文的基本内涵与特色，要求言简意赅，核心明确。

### 1.1　英文标题要求

（1）确切、简炼、醒目（Correct, Concise, Clear）。确切，即能够准确表达并切合本文的实质内容；简炼，即以最少的文字概括尽可能多的必要内容；醒目，即要一目了然地反映论文特色，甚至要有一定的生动性和新颖性。避免文章标题过繁或过泛。

（2）中、英文标题内容应一致。原则上同一篇论文的英文标题应与中文标题内容一致，标题的翻译应力求主题突出，开门见山，而不是逐词对应直译，允许非实质性词不同。语法要严谨，形式要新颖，并应该服从英语的语言习惯，如中文题名中的“体会”、“探索”、“研究”、“初探”、“论”、“讨论”等词，在英文标题中尽可能省略，即“The Effect of …”、“Study of …”、“On …”、“A Discussion on …”、“Research on”、“A Preliminary Study of …”、“Preliminary Discussion of …”等这一类的词尽可能省略。这些套式省略既符合英语标题的潮流，使标题更加简洁醒目，主题突出，又符合英语的表达习惯和国际刊物的基本要求。因为英文的“Preliminary Study (Research /Discussion /Investigation) of …”在中国人看来是一种谦虚的写法，而给英美人很容易造成一种错觉，认为研究不深透，处在初步或初级阶段，不宜形成论文。

### 1.2　标题的构成及形式

标题一般由短语构成，通常采用名词性短语（noun phrase）；一般不用完整的陈述句；偶而使用探讨性语气的疑问句作标题。

标题的形式大致可分为三类：①以突出研究目的或研究对象为主的；②以突出研究结果或研究方法为主的；③将研究对象、研究目的和研究结果兼而有之进行表达的。但也可不受此限制。

### 1.3　副标题的处理

一般限制在简要的正题之后加副题的题名，只有必要时才使用。要求主题与副题之间用句号分开。副标题的翻译一般分以下三种情况。

（1）当副标题非常重要并且是标题的有机组成部分时，应全部译出。

（2）从英语角度看，副标题比主标题更重要时，可互换位置。

（3）汉语标题内容较多，或为了强调某些词，译成英语时可把标题分为主标题和副标题，使重点突出。

**1.4 标题的其他要求**

（1）标题中的第一个词和每个实词（例如，名词、动词、形容词和副词等）以及4个以上（含4个）字母的连词和介词等首字母要大写，而含3个或3个字母以下的虚词（如连词、介词等）全部小写。例如：

限量灌溉对小麦抗旱增产和水分利用的影响

Effects of Limited Irrigation Upon Yield Increase in Wheat by Drought Resistance and Water Use.

拖拉机履带板材料的研究及磨损试验

Material for the Track Shoes of Tractor and the Abrasion Tests.

（2）有的刊物要求标题的所有字母都采用大写。例如：

应用仿真技术估算模型参数的方法

A METHOD FOR ESTIMATION MODEL PARAMETERS BY SIMULATION TECHNIQUE。

（3）标题中可用可不用的冠词（包括不定冠词和定冠词）一般可以省略。

例如：《环氧树脂——玻璃粘接在水中的破坏》

(The) Environmental Failure of Epoxy Resin—Glass Joint in Water.

（4）限制使用缩略语（abbreviation），只有那些全称较长、繁冗，而且其缩略语已被科技界公认，为广大读者中所非常熟悉的，才可使用。缩略语全部用大写字母，如：LASER，AIDS，CAD，GIS 等。

（5）特殊字符，如数学符号和希腊字母在题名中尽量不用。

## 2 作者姓名及单位的翻译

作者姓名按汉语拼音写全称，一般姓（surname）在前，名（given name）在后，姓和名的首字母均大写，当名的第二个汉字为零声母时，在该拼音前加隔音符号“'”。例如：Guo Xing'an (郭兴安)。有的刊物要求姓的所有字母和名的首字母都大写，如 YANG Deguang。也有的刊物按英美习惯，名在前，姓在后。如果作者不止一人，可翻译第一作者，其后加上 et al (“等”)，如，Wang Dianwu, et al. 也可将所有作者都译出。每个作者下一行注出其单位，同一个单位的几个作者只注一个单位；单位用英译名；作者单位后给出其所在城市的汉语拼音和邮政编码。如作者为西安理工大学，可译为：Xi'an University of Technology, Xi'an 710048

## 3 摘要（Abstract）

**3.1 摘要的类型**

试验研究与专题论述的科技论文要求写情报性摘要；综述、评述性论文论题不集中而篇幅较长的，写指示性摘要。其各自要求及内涵如下：

（1）情报性摘要（informative abstract）：概括地陈述研究目的、方法、结果与结论等，尽量多地提供文中定量和定性的信息。

（2）指示性摘要（indicative abstract）：简要介绍文章的性质、主要内容（论题）、阐述方式及取得的进度，一般不用提供定量数据。

### 3.2 摘要的要素

（1）目的(purpose)：准确描述该研究的目的，若有多个研究目的应择其要者加以阐述。

（2）背景(background)：对该方面的研究起源、发展和研究现状作一概要描述。写明出于何种考虑，为了解决什么问题，在何种条件和情况下进行的研究。

（3）方法（methodology）：描述该研究所用的仪器设备、研究对象和方案以及所用原理、定理，对无试验内容的论文，要说明数据来源、对数据的处理方法及效果评价标准等。

（4）结果（results）：言之有物地描述所得结果，可以是试验性的或理论性的结果，或者是所收集到的数据，也可以是记录下的相关关系以及观察到的效果和性能、记实、推理及论证结果等。要说明各数据是首次得到的还是推导得出的，它们是一次观测还是重复测量所得的结果。当所得数据太多而不能全部纳入摘要时，则应优先考虑以下结果：经过验证的新事件，长期数据，有意义的新发现，与前人理论相矛盾的结果或作者所提出的结果与某实际问题有关。应当指出准确性和可靠性的界限和适用范围。

（5）结论（conclusion）：有时是结论及讨论（conclusion and discussion），即对结果的分析、比较、评价及其应用等，也包括所提出的，今后应进一步研究的课题、建议、预测，所确立的新的规律性（关系）和假设的成立、修改和推翻等。注意这里是指分析评价其应用范围、精度等客观事实，切忌对论文做主观评价。即摘要中不允许写"…具有重要意义"、"可观的推广应用前景"、"为…提供依据"等自我评价语言。

### 3.3 英文摘要的结构格式

上述要素对中、英文摘要都一样。但英文摘要的结构格式（要素排列），除传统的与中文结构格式相同外，更多地采用开门见山的结构格式，即主要结果、结论，支持性（证明性）细节和方法，具体结果及其他信息等。摘要传统结构格式举例如下：

Effects of Unsmoothed Surfaces on Soil Adhesion and Draft Bulldozing Plates

The draft of unsmoothed surfaces of bulldozing plates decreases with increases of soil moisture contents. An average maximum reduction reaches 30% in draft (Major results and conclusions). This study designed a number of unsmoothed surface morphology of heads of dung beetle. Draft of these plates varied with the design /distribution of small convexes glued on the surfaces of the bulldozing plates, material of the small convexes, speed of cut, angle of cut, depth of cut, and soil moisture content. The tests covered the range of soil moisture content from 30.38%(db) to 35.5%(db) (methodology, supporting detail). The experiment results also indicate that Ultra High Molecular Weight-Polyethylene can reduce soil adhesion and draft considerable (other findings).

### 3.4 英文摘要的语法

（1）时态（tense）：用一般过去时叙述研究过程（包括实验、观测、数据处理等）；用一般现在时阐述研究结果、结论、叙述客观事实等，表示该文"报告"、"描述"及"讨论"等意思时也用一般现在时。一般不用完成时、进行时和其他复合时态。

（2）语态（voice）：谓语动词尽量使用主动语态，少用被动语态；但在叙述研究过程时可用被动语态使重点突出。例如：this paper presents formulas 和 formulas are presented，前者简洁、有力，后者突出重点。

（3）代词（pronoun）：以客观事实作主语，尽量不用人称代词，禁止使用第一、第二人称代词。必要时用动名词作主语。

（4）句型（sentence）：尽量用简单句来表达，即使有必要使用复合句型，也要尽量使用简单复合句，有些从句可改用分词短语。

## 3.5 英文摘要的措词

（1）准确使用近义词。要仔细辨别词义相近的词汇，选择最准确的词汇。

（2）准确使用科技规范词。英语中的科技词汇都有其独特的表达习惯和固定用法，特别是科技名词述语和专业述语。要求尽量从正式出版的专业词汇表和权威专业词典中查找科技词汇、技术名词，专业述语要用固定用法，规范表达。

## 3.6 英文摘要的行文方法

### 3.6.1 研究目的的表达方法

第一句应直接点题，可以有几种写法。

（1）主动句型

This paper describes / presents / discusses / analyses / reports on / investigates / examines / deals with / researches into / gives / points out / reviews / … 本文描述/提出/讨论/分析/报告/调查/检验/论述/探讨/给出/指出/总结/…

This paper is concerned with / aimed at / limited to / related to … 本文研究/旨在/限于/关于…

The purpose (aim/objective / …) of this paper is to discuss /study / research / … 本文目的是…讨论/研究/探讨/…

也可以使用名词句型：

The paper makes a study of …

This paper makes investigations on …

也可不用 this paper 作主语，而用 author（作者）作主语，但一般不用 I 或 we 作主语。有时还可将 this paper (author)的主语省略，而直接采用 discusses/presents / deals with.

（2）被动句型。

Information regarding (或 concerning) …is described / presented / discussed / analyzed / reported / investigated / examined / dealt with / given / pointed out.

In this paper a new method (或 approach) of …is introduced (或 recommended). 本文介绍的一种…的新方法。

By using …，…was studied. 使用…，对…进行了研究。

### 3.6.2 研究方法的表达方法

在第一句之后便开始简述过程或论据，其句型有以下几种方式。

（1）叙述理论研究、分析方法或设计方法，如：

Use is being made of the concept of … 正在使用…概念

Conditions are considered for / of … 考虑了…的情况（条件）

The approach is based on … 这种方法以…为基础（建立在…基础之上）

The requirements for …are noted 这注意到有关…必要条件

（2）叙述使用数学模型的情况，如：

Patterns of …are studied 对…模型进行研究

The formula is derived for …according to … 按照…导出了…公式

（3）介绍实验或调查的情况，如

Tests have been carried out to study … 进行了试验，以研究…

The experiment is performed using … 用…做实验…

Cross has been made using … 用…进行杂交…

Test methods are listed. 试验方法已逐一说明。

An experimental investigation is described in this paper. 本文叙述了一项实验调查。

以上 3 种情况也可归纳用以下几种方式开始。

An example of …is analyzed / described / discussed / examined / studied in detail. 详细地分析/描述/讨论/检验/研究/了…的例子。

Data are displayed in graphs and tables. 数据显示在图表中。

Findings (或 Results)are presented / reported / analyzed / examined / discussed. 提出/报告/分析/检验/讨论了新发现（或结果）。

A series of experiments were made / carried out on … 对…进行了一系列实验。

Special mention is given here to … 这里专门提到…

Examples of …demonstrate that … …的例子表明…

Statistics confirm … 统计数字肯定了…

### 3.6.3 研究结果与结论的表达方法

对这一部分的内容所采用的句型一般有：

Facts show that … 事实证明…

Experiment finds that … 实验表明…

Study proves that … 研究证明…

Results show / indicate / reveal /suggest / illustrate / demonstrate / … 结果证明/表明/揭示/提出/说明/表明…

Comparison concludes that … 这一比较推断出…

Statistical analysis demonstrates that … 统计分析提出…

根据摘要内容的不同，还可具体分为以下几种表达方法：

（1）说明理论、分析方法或设计方法的行文句型。

The result of this study can be generalized / finalized for … 这一研究成果可推广到/概括…

Acceptable results of design were obtained by the method for … 通过…方法，获得了在设计上可以接受的结果

The results of this study are summarized / summed up as follows / in the following. 研究结果归纳如下。

（2）说明使用数学模型和公式结果的表示方法。

Calculations made with this formulation show that … 用这一公式计算表明…

An exact expression is obtained and the results are analyzed. 获得了一项正确公式，并且分析了其结果。

（3）说明实验或其他方面结果的表达方法。

The conclusions were drawn from the test results. 从试验结果引出（得出）这些结论。

Results for … are found to be close to the experimental data.　已经证明…的结果与实验数据相接近。

The results are illustrated by a specific example.　其结果可由一具体实例加以说明。

#### 3.6.4　摘要的末尾句表达方法

These results / data /findings /experiments also indicate that …　这些结果/数据/发现/实验/已指出…

The findings imply that …　这些发现暗示…

Based on these conclusions …is discussed　根据这些结论，讨论了…

The findings suggest that further research into …is called for (或 would be worthwhile)　这些发现提示对…应进一步研究（或对…进一步研究是值得的）

### 3.7　英文摘要写作的其他注意事项

摘要具有独立性，简炼且完整，是一篇高度浓缩的不依附原文的短文。英文摘要不仅要结构合理规范、内容丰富完整，而且要符合英文摘要的特殊要求和科技英语的表达习惯。篇幅以200～300个实词为宜。除措词确切外，还应特别注意句子表达要简洁精炼。注意事项如下：

（1）文摘第一句切不可重复题名。

（2）句子结构要严谨完整，尽量用简单句，但不能不完整（不能用电报语言）。

（3）删繁就简，能用一个词的不要用词组，能用简单词的不要用复杂词。如，用 because 不用 due to，用 to 而不用 in order to，用 use 而不用 utilize 等。

（4）可用动词的情况尽量不用其名词形式，避免使用 be, have, do 等弱动词。如，用 thickness of plastic sheets was measured，而不用 measurement of thickness of plastic sheets was made.

（5）避免使用长系列形容词或名词来堆列修饰名词，可用预置短语分开，或用连字符断开名词词组，作为单位形容词（一个组合形容词）。如 water saving irrigation agriculture 改用 agriculture of water-saving irrigation。

（6）文词要朴实无华，不用文学性描述。尽量简化一些措词和重复的单元。如：

| 不用 | 而用 |
| --- | --- |
| at a temperature of 250℃to 300℃ | at 250～300℃ |
| at a high pressure of 200 Pa | at 200 Pa |
| at a high temperature of 1500℃ | at 1500℃ |

（7）组织好句子，使动词尽量靠近主语；用重要的事实开头，尽量避免用辅助从句开头。

（8）用标准英语。可用英式或美式英语拼写，但每篇应保持一致。

（9）缩写词第一次出现时必须用括号注出其全称，公认者除外。

（10）限制使用特殊符号（数学符号、希文字母等）、数学表达式和图表。

## 4　关键词（Key words）

一般从论文中选择3～8个单词或短语作为关键词，表示论文中的信息，读者可以从关键词了解其论文的内容，同时也为文献检索提供方便。关键词排在摘要的左下方，各词之间用逗号“,”或分号“;”隔开（根据刊物要求选用）。

关键词尽量选用主题词。对主题词表中未列出的新技术专业名词及词组，可自由选定，作为自由词，用作关键词。

# 参 考 译 文

## 第一课 水 的 重 要 性

在地球表面以相对纯的形式存在的一切化合物中，水是人们最熟悉的、最丰富的一种化合物。在水中，氧这种丰富的化学元素与氢结合，其含量多达 89%。水覆盖了地球表面的大约 3/4 的面积，并充满了陆地上的许多裂缝。地球的两极被大量的冰所覆盖，同时大气也挟带有占其重量 0.1%～2%的水蒸气。据估计，在温暖的夏日，每平方英里陆地上空大气中的水量约为 5 万吨。

地球上所有的生命都有赖于水，水是活细胞的基本要素。人类、植物和动物都得用水。没有水就没有生命。每一种生物都需要水。人可以接近两个月不吃食物而仍能活着，但不喝水则只能活三四天。

在我们的家庭中，无论在城市还是农村，水对于卫生和健康来说都是必不可少的。美国家庭的年平均用水量达 6.5 万～7.5 万加仑。

水可以被认为是最基本的和最廉价的原料。我们的农产品，大部分都是由它构成的。水是农作物和动物生长的要素，也是奶类和蛋类生产的一个很重要的因素。动物和家禽，如果用流动的水来喂养，那么每磅饲料和每个劳动小时会生产出更多的肉、奶和蛋。例如，苹果含 87%的水分，苹果树就必须吸收比苹果重许多倍的水分；土豆含 75%的水分，那么种植每英亩土豆就需要若干吨水；鱼体内含 80%的水分，鱼类不仅要消耗水，而且还必须有大量的水才能在其中生存；牛奶含水量为 88%，为了生产每夸脱牛奶，母牛需要 3.5～5.5 夸脱的水；牛肉含 77%的水，为生产 1 磅牛肉牛必须饮用许多磅水。如果缺水，就会使农产品减产，就像缺乏钢会引起汽车产量下降一样。

水除了直接为我们的家庭和农场利用之外，它还以许多间接的方式对我们的生活产生影响。在制造、发电、运输、娱乐以及其他许多行业，水都起着很重要的作用。

我们对水的利用随人口的增长而迅速增加。在许多地方，无论地面水或地下水都已经严重缺乏了。由于任意污染河流、湖泊和地下水源，已经大大地损害了人们能够利用的水的水质。因此，人人有责对水采取保护措施和卫生措施，这对于我们人类的未来是极端重要的。

## 第二课 水 循 环

在自然界中，水总是不断地从一种状态改变成另一种状态。太阳热使陆地和水面上的水蒸发。这些水蒸气（一种气体）由于比空气轻，会上升直至达到高空冷气层，并在那里凝结成云。云层随风飘荡，直至遇到更冷的大气层为止。此时水便进一步冷凝，并以雨、雹或雪的形式落到地面。这样便完成了水的循环。

然而，完整的水循环要复杂得多。由于蒸发作用，大气不仅从海洋而且从湖泊、河流和其他水体，以及从湿的地表面获得水蒸气。也可从雪地中雪的升华和从植物与树木的蒸腾获得水蒸气。

降水可有各种不同的出路。大部分降水都直接落到海洋。落在陆地区域的水，有些被植物所摄取，或在降至地面之前就蒸发了；有些被封冻在雪原或冰川中达一个季度乃至成千上万年；有些则因储存在水库、土壤、化合物以及动植物体内而滞留下来。

降到陆地区域的水可能作为溪流与江河的径流，或在温暖季节融化的雪水直接回到海洋。当降水不立即流走时，它会渗入土壤。这些地下水中有一些被植物的根吸收，有一些则通过下层土壤流入河流、湖泊和海洋。

因为水对于维持生命来说绝对必要，在工业上也很重要，所以人们为了自身的利益试图以各种方式来控制水的循环。一个明显的例子就是在一年中不同的时间根据当地降水的多寡（按对水的需要来说）将水储存在水库中。另一种方法是试图将干冰或碘化银微粒射入云层来增多或减少天然降雨量。虽然这种改造气候的方法迄今只取得了有限的成功，但许多气象学家都认为，有效地控制降水在将来是可以做得到的。

其他一些影响水循环的努力包括沿等高线耕作梯田，以使径流减速，让更多的水渗入地下；建筑堤坝以防洪水等。在水回归大海之前将它重复使用，也是一种常用的办法。自河道取水的各种供水系统可将水在最终到达河口之前，经过净化，可重复使用多次。

人们还试图预测水循环过程中一些事件的结果。例如，气象学家预报一个流域的降雨量和降雨强度；水文学家预报径流量等。

# 第三课 水 文 学

## 1 历史

最早的水利工程在有史以前已消踪匿迹了。也许史前的人曾发现横贯河流的一堆石头就能提高水位，足以淹没作为生长野生食用植物源泉的土地，而这样在干旱季节就能给植物浇水。不论水力学的早期历史如何，充分的迹象表明，建造者们还不懂得多少水文学知识。早期的希腊和罗马文献说明这些人承认海洋是一切水的主要源泉，但是不能想象降雨量会等于或超过河道径流量。当时典型的想法是海水从地下流到山脉底部，那儿有一个天然蒸馏器除去水中的盐分，水汽通过管道上升到山顶，在那里凝结，并从河流的源头流走。M.V.波利欧（大约公元前100 年）看来就是像我们今天这样认识降水作用最早的人。

L.达芬奇（1452—1519）是提出水文循环现代观点的第二个人，但一直到 P.贝罗特（1608—1680）才把观测的雨量与估算的塞纳河的径流量进行比较，说明河川径流量约为降水量的1/6。英国天文家哈雷（1656—1742）从一个小盘子中测得蒸发量，并且用这一资料估算地中海的蒸发量。然而直到 1921 年，有一些人仍然对水文循环的概念表示怀疑。

印度早在公元前 4 世纪就测量降水量了，但是令人满意的测量河道流量的方法很迟才得到发展。公元 97 年，罗马水利专员福朗堤努斯只按横断面面积估计流量，而不考虑流速。在美国，有组织地测量降水量是 1819 年在陆军军医总监领导下开始的，1870 年移交给通讯兵团，最后，在 1891 年移交给新改组的美国气象局，该局于 1970 年改名为国家气象局。早在 1848年密西西比河上就进行分散的河道流量测量了，但是，直到 1888 年美国地质调查局承担这项工作时，才开始实施系统的观测计划。霍德、米德和谢尔曼等人在 20 世纪早期刚开始对这一领域进行探索，因此，在这时期之前，在水文方面没有进行什么定量工作是不足为奇的。大约从 1930 年起，由于在防洪、灌溉、土地改良和有关领域中开展了大量活动，第一次为有组织地研究水文学提供了真正动力，因为需要更精确的设计资料，这已是十分明显的事了。大多数现代水文学的概念从 1930 年就开始有了。

**2 水文学在工程中的作用**

在工程上，水文学主要用于水工建筑物的设计和运行。溢洪道、公路涵洞、或者城市排水系统会期望有什么样的洪水流量？需要多大的水库库容才能保证干旱季节里有足够的灌溉水量或城市供水呢？水库、堤坝或其他控制工程对河流洪水流量有什么影响？这些典型的问题等待水文学家去解答。

像联邦和州辖水利机构这样的大型组织，拥有一批水文专家来分析他们的问题，但较小的单位往往没有足够的水文工作给专职水文专家做。因此，许多土木工程师们应邀进行临时的水文研究。这些土木工程师处理的工程和年费用可能比水文专家还多。无论如何，水文学的基础知识看来是培训土木工程师所必不可少的一部分。

**3 水文学研究的主要内容**

水文学研究许多问题。本书所介绍的主要内容可大致分成两个方面：收集资料和分析方法。2～6 章研究水文学的基本资料。充足的基本资料是任何一门科学所不可少的，水文学也不例外。事实上，水文现象中包含着许多自然过程的复杂特征，用严密的推论来处理许多水文过程是很困难的。人们并不总是能够从基本的自然法则出发，并由此来推求预期的水文结果。相反，从大量观察的事实出发，分析这些事实，并根据分析建立控制这些事件的系统模型却是十分必要的。因此，对于没有足够历史资料的特殊疑难地区，水文学家就将陷入困境。大多数国家有一个或更多的政府机构负责收集资料，重要的是要让学生学会这些资料是如何收集和刊出的，了解这些资料精确度的局限性，学会整理分析和校正这些资料的专门的方法。

典型的水文问题包括估算小的数据样本中无法观测到的极值及估算无资料地区（这种地区比有资料的地区多得多）的水文特征值，或者估算人类活动对该地区水文特征值的影响。一般来说，每一个水文问题都是不同的，因为它涉及到特定流域内特有的自然条件。因此，某种分析所得的定量结论常常不能直接移用到另一个问题上。然而，应用一些比较基本的概念可以得出大多数问题都适用的一般解决方法。

# 第四课 地 下 水

地球上的总水量中，97%在海洋，2%在冰川中，只有 1%在陆地上。陆地上的水几乎全部（97%）埋藏在地面以下，称为地下水。大部分地下水或通过地下流动，回到海洋；或先进入河流和湖泊，最终又回归海洋。

这些广阔的地下含水层为干旱地区和灌溉区域提供了迫切需要的水分。地下水的作用与地表水的作用相似，也以均夷作用塑造着地貌。

尽管人类自古以来就知道地下水，但对它的特性、发生、运动和地貌意义还不清楚。然而，近来关于地下水与水文循环关系的这些错综复杂的问题已找到了一些答案。

**1 地下水的来源**

自从公元维特鲁维亚时代以来，为了解释在地面以下存在大量的地下水，已经提出了许多理论。一种理论认为：只有海洋能提供大量的地下水，地下运动的水来自海岸带。维特鲁维亚是第一个认识到降水是地下水的主要来源，尽管他对所涉及的力学方面的解释很不科学。

现在，他的理论已经确立，称作渗透理论：认为地下水是水从地表渗入到地下的结果，或者直接来自降雨，或者间接来自河流和湖泊。这种水称作天落水。地下水总量中的很小一部分来自其他水源。原生水是在沉积岩形成时滞留在其中的水。岩浆水是由于在相当深处的地壳运动对地壳所增添的水，火山水就是一个例子。

## 2 地下水的分布

在降雨期间，水渗入地下。在重力的影响下，这种水通过土壤颗粒间的孔隙向下流动直至水流到达不透水岩层为止。向下移动的过量水分，充满了土壤颗粒之间的孔隙，挤出了土壤中的空气。在雨水过多时，整个土壤剖面达到饱和状态，而在干旱时期就不存在饱和土壤。通常，饱和土壤的上部界限称为地下水位，约在地面以下 1 米左右，其高度取决于土壤特性和降雨量。

根据水充满孔隙的程度，地下水分成两层：包气带和饱水带，如图 4.1 所示。

### 2.1 包气带

这一区域从地面向下延伸到水面——地下水位。根据该层中所含水的形成和运行情况，可进一步划为三个带：土壤水分带、中间带和毛细饱和带（图 4.1）。

（1）地壤水分带。假设土壤是干燥的，初期降雨量就向土壤中入渗，入渗量取决于土壤结构。主要由大颗粒构成的土壤，在每个颗粒之间有很大的孔隙，在正常情况下入渗速度要比在小颗粒组成的土壤中快得多。不论土壤由什么构成，土壤颗粒表面由于分子引力吸附着一些水，就形成一层水膜，阻止重力水向下运动。以这种方式保持的水称为吸湿水。即使它不受重力的作用，它也能蒸发，但是通常不能被植物吸收。

（2）中间带。中间带发生在干旱季节，地下水位在地面以下相当深的地方。中间带与土壤水分带相似，也是通过分子引力把水分保持在土壤颗粒上。但差别在于：中间带的水膜不能通过蒸腾或蒸发回到大气层。在湿润地区，降雨相当可观，中间带可能不存在或者非常浅。重力水即渗漏水通过中间带向下渗漏到饱水带。

（3）毛细饱和带。这一带位于地下水位以上，是个很浅的含水带，这些水是由毛管力从地下水库吸引上来的。这一层的深度完全取决于土壤的结构，含有微小孔隙的土壤比含有大孔隙的土壤能从下面吸取更多的水。在后一种类型的土壤中，分子引力不能跨越土壤颗粒间的空隙。因此，砂质土很少出现大面积的毛细饱和带，而土壤水分带和饱水直接相通。

### 2.2 饱水带

饱水带是孔隙完全被水充满、没有一点空气的土石层。这一层的技术术语叫做地下水，尽管这一术语广义上也包含包气带的水。饱水带的上部边界是地下水位或地下水面。要了解地下水层的深度是困难的，虽然大多数地下水是在地壳表层 3 千米以内发现的，但是能够保存水的微小孔隙可以延伸到 16 千米的深度。这似乎就是贮水岩层的上限。那里的压力大到足以充满任何孔隙。

在不同的地方挖探测井就能完全标绘出饱和层的上界面。通过研究提出了两个十分有趣的论点（图 4.2）。

1）在地面最高的地方，地下水位也最高；而在地面最低的地方，地下水位就最低。丘陵和山脉的地下水位比峡谷和湖泊的地下水位高。这是因为水通过包气带不断地渗透提高了地下水位；地下水层的渗流进入河沟和湖泊，从而降低了地下水位。

2）在山区高地，地下水位的埋藏深度最大，在重力作用下，那里的地下水非常顺畅地向下流动。靠近河流、湖泊和沼泽地区时，地下水位即使未达到地表，也接近地表，因为水从较高的地方汇集而来使地下水位升高。

# 第五课　洪水是怎样形成的

洪水是怎样形成的？根本的原因是流域内过多的径流汇入河系，超过了河系的承载能力。科学技术能防止洪水发生吗？或者至少能降低洪水的危害吗？遗憾的是，这是一个迄今还没有满意答案的复杂问题。

让我们首先考虑一下如何减少流域的径流。有些地区的土壤吸水能力很低。下暴雨时，这种土壤很快就饱和了，多余的雨水全部流入河道。季节性的变量就是暴雨开始时土壤的水分状态。如果土壤已经湿润，较小的暴雨也会引起大量的径流，因为这种土壤已不能保持更多的水分。这些因素不易受人类的影响。然而，人类对流域土地的利用情况会对洪水产生重要影响。大规模地砍伐树木会降低土壤保持水分的能力。砍伐树木也会加剧土壤侵蚀，导致河道淤塞，从而加剧了洪水。所以对流域正确开发利用是控制洪水的一项重要方法。

筑堤是在紧急情况下可采用的更加直接的方法。当上涨的洪水威胁城镇时，居民通常组成修堤队，沿河岸用砂袋筑堤，想用岸堤挡住洪水，直到紧急状态消失。人们或许想知道为什么通常不把河堤建成永久性设施，让城镇永保安宁。原因在于筑堤不是解决问题的圆满方法。如果堤岸崩溃，洪水涌出，就成为破坏能力剧增的突发性水灾。大堤会使一个区域来的洪水转向，常常在另一地区引起问题或使问题变得更为严重。因此大堤也会成为居民点之间不和的原因。

另一种方法是筑坝，让洪水存蓄在水库里，而在随后的几周或几个月内把洪水慢慢地放掉以度过危机。灌溉和防洪两用的坝似乎是合理的解决方法。遗憾的是，用于灌溉的水库冬季要蓄水，而用于防洪的水库却要放空，以便需要时，可以蓄水 。运行要求的矛盾意味着两用坝很少能办得到。防洪需要单独筑坝，造价很高，这又使它成为不切实际的解决办法。能解决这一问题的另一个办法是改善河流的过水能力，使河流能宣泄更多的水量而两岸并不漫溢。有许多措施是可行的，有的简单，有的复杂。所有这些方法对河流都普遍有效。所以任何一种措施都应作为全盘计划的一部分。这种工作称作“河道整治”或“河道管理”。

一种简单而重要的方法是确保河道不被阻塞。阻塞物常为倾倒在河内的枯树，一直阻碍水流。这些阻塞物称为“水中隐树（沉木）”；而清除工作称为“清除隐患（沉木）”。在澳大利亚，河流两岸的许多树木都是硬木。树木太重浮不起来，所以就一直留在树木倒下去的地方。此外，硬木经久不烂，如大红枫树原木，在水下经过了 100 多年还保持完好。

还有一种增加河道过水能力的方法是清除阻塞河道的植物。早期的移民在河岸上栽插了许多柳树。这样做的部分原因是为了遮阴，部分原因是为了回忆古老的英格兰 ，同时也希望减少河岸侵蚀。遗憾的是，这些树很难控制，柳树蔓延现已成为普遍的难题。保护河岸免受水流的侵蚀也是一种措施。弯曲河道的凹岸总是遭受侵蚀，这意味着可贵的土壤因河岸侵蚀而流失，也是当地洪水的一种起因。人们已经想出多种保护河岸免受侵蚀的方法。

最简单的护岸方法是沿着受侵蚀的河岸固定或系上树干。树干能保护河岸，促使淤泥在河岸沉积，使被侵蚀的河岸逐渐得到修复。

## 第六课 水污染的性质

水是人类最宝贵的资源之一。但在因水质降低或水源不足而危及水的利用之前，人们一般都不珍惜它。水的污染主要是由于人类的活动，特别是人类对水源管理不当造成的。污染物指一切影响水的自然状况或预期用途的化学的、物理的或生物的物质。因为水的污染影响了水的供应、水质及用途，所以它引起了世界范围的极大关注。

随着全世界用水量及用水类别的增加，已产生了各类必须满足的水质标准要求。这些要求包括：①保持河流的自然状态；②供作饮用水源；③保护与繁殖鱼类及野生生物；④农业利用的安全；⑤娱乐利用（包括游泳）的安全；⑥适合各式各样的工业用途；⑦无有害物；⑧公用发电；⑨稀释和运输废物。除了为满足上述多种用途所需的化学、生物学及物理学上的特定要求外，还有一些反映公共卫生要求、美学、经济学以及短期与长期生态学影响方面的限制。因为构成水污染的定义必须考虑水的预期用途，所以水的污染并没有严格的或明确的定义。

区分构成水污染的气态、液态和固态组分的一种方法，是根据水的预定用途。这样，污染物便分为不允许的、不符合要求或不适宜的、可允许但不一定理想的、或符合要求的。例如，如果水预定用作动物消耗，那么有毒化合物就不符合要求，而有一定的含氧量则是适宜的。另一方面，如果水计划用做火力发电厂生产蒸汽，则毒物可能是允许的，甚至或许还是符合要求的，而可能腐蚀设备的氧则成为不适宜的。

作为人类生活、工业及其他活动的结果而进入水中的污染物，其分类的另一种方法是区分成永久性或非永久性污染物。永久性污染物是不被天然水体中的生物过程所改变的那些污染物。这些污染物大多是无机化合物，它们在受纳水体中被稀释，但其总量不会有明显的改变。工业废水含有大量的这类污染物，包括金属盐类、其他有毒物、腐蚀性物质、色素和引起异味的物质。家庭生活污水以及从农灌的排水也可能含有大量的这类污染物，包括氯化物和硝酸盐。

另一方面，非永久性污染物则可以由于水体中生物现象所引起的化学和物理过程而改变其形态或降低其数量。最普通的非永久性污染物的来源是家庭生活污水，这是一种高腐化性有机废水，它能由于水体中细菌和其他微生物的作用而转变成为无机物。例如，转变成重碳酸盐、硫酸盐和磷酸盐。

如果水体并没有接受太多的污水，它就可以“自净”。这个过程包括好氧细菌的作用，即需游离氧的细菌分解污染物的作用，并且还不会产生讨厌的臭味。

然而，如果水体接受的污水超过一定数量，那么这种生物降解过程就会变为厌氧过程，即不需要游离氧的细菌作用来完成。在这一过程中，会产生有害的硫化氢气体、甲烷气体和其他气体。天然河道中发生的这些好氧和厌氧过程被用于污水处理厂中，两者实际是污水处理的主要单元过程。

水的污染问题几乎一直是并且现在仍然是世界范围的问题。

## 第七课 水资源开发规划

规划就是对一个工程从目的的初始陈述、方案评价到行动过程的最终决定的有序研究。

除了建筑物的细部工程外，它包括与工程设计有关的全部工作。它是决定一个建设项目继续进行（或放弃）的依据，也是这个项目工程的最重要方面。因为每个水开发项目对经济和自然的调节都是惟一的，所以，一个简单的描述方法不可能获得最好的决定。在项目规划途径的方法选择中，没有什么能够代替“工程判断”。但朝着最终决定的每一独立步骤，无论何时都应采用定量分析的方法，而尽可能不用估计或判断。人们经常听到“流域规划”这个词，但规划阶段对最小工程来说是不太重要的，整个流域的规划比单个工程涉及更加复杂的规划程序。而单个工程在取得正确决定时所面临的困难可能与整个流域规划一样大。

“规划”这个词包含有不同于上面提到过的另外的涵义，这是一个区域总体规划的概念。它试图为一个地区确定最适宜的未来发展模式。如果这个总体规划就是实际上的最适宜的发展模式，那么未来的发展就应该向这个模式方面引导。遗憾的是“最适宜的”概念是主观的，它很难保证任何总体规划初次被提出来就能满足这一高标准。其次，技术的变化、经济的发展和公众态度常常使一个总体规划在一个相当短的时间内就变的过时。任何一个规划都是基于有关未来的假设，因此，如果这些假设没有实现，那么这个规划就必须修改。规划也必须定期进行全面的修改。

一个经过细心研究并与其他区域规划紧密协调的区域水管理总体规划，可能是一个有用的工具，在决定许多可能方案时应当采用这一有用工具。但必须考虑技术、经济和社会环境的变化或当研究出新的有关数据时，随时对它进行调整。一个总体规划，只不过是所有可行性研究中的一种，很可能被证明是没有价值的。

每一个国家许多领域都有规划，而每一领域中规划工作的目的和性质又不相同。许多国家有国家规划机构，它们的目标是加强国内的经济发展，改善国内的社会状况。即使没有这样的组织存在，也有国家目标存在，国家规划的某些形式由立法机关或政府行政部门提出。国家规划机构很少直接处理水问题，但在调整粮食、能源、工业商品和住房等生产指标时，它可能有效的为水管理规定目标。

考虑到一个国家各地区之间存在着差异，因此，一个国家可以有地区规划分支机构，并且可以存在相应的地区水规划。然而，由于水规划的一个自然“区域”是流域，因此有建立流域委员会或流域管理局（官方）的趋势。这些规划组织必须确保流域内各项工作之间的协调。水管理的每个具体行动可能对下游（并且有时对上游）产生影响。因此，这些具体行动不应单独进行规划，而必须进行相互之间的协调。

具体行动计划是规划的最低级，它决定了水管理有效性的实现。但决定水管理效率的重要决定就是由这一级制定的。尽管一个具体工程不一定产生，但这一级别的规划常被称为工程规划。例如，（因）一项研究而产生的一个洪泛平原管理方案就是一个合法的工程规划。

工程规划通常要经过几个阶段才能成为最终的方案。在每一个国家，规划都有一个具体名称的规定顺序。第一阶段或设计任务书阶段，通常是一个粗过滤网，这一阶段是用来淘汰那些具有明显不必广泛研究特征的工程或方案，从而确定哪些是值得进一步研究的方案。设计任务书阶段后可能有一个或更多的阶段要对建议方案的可行性进行详细的评价。在这个过程中，要对最适宜方案的作用进行系统说明，这就是计划。在许多情况下，一个单独的可行性分析是足以满足要求的，因为方案的性质比较容易评价。在其他情况下，一个或更多的初步可行性分析可以对建议方案的各个方面进行检查。几个阶段研究的目的，首先是通过考查工程的最弱方面来减少规划的费用。如果这个工程由于某方面的原因被淘汰，那么这个工程所有其他方面的研究费用就不用再花了。如果这些分析序列越来越多地成为这个项目所有各方面的一系列的更详细的观点，那么规划费用就会因许多事情将被重做两次或更多次而增加。

可行性研究通常要求一个工程的结构细节应该规定得充分具体的，以便给出一个准确的费用估计。在规划工作的最终阶段，设计的细节必须进行认真的审查，并且要给出施工图和设计说明书。尽管在原则上可行性分析的结果取决于可行性研究的基础，但更详细的研究可能开发出使可行性改变的资料的可能性总是存在的。所以，不到最终设计完成就不能取消继续做下去的决定。

# 第八课 水 库

拦河筑一条象坝的障碍时，水就被拦蓄在障碍物的上游并形成水塘，通常称之为水库。

总而言之，被蓄集在水塘或湖泊里的水就称为水库。水库所蓄的水可有多种用途。水库根据其用途可分为以下几类：①蓄水水库；②防洪水库；③配水水库；④综合利用水库。

## 1 蓄水水库

城市给水、灌溉供水或直接从河流中取水的水电工程或者溪流，在枯水期，也许不能满足用户的需要；在丰水期，由于洪水泛滥它们又可能无法发挥作用。蓄水水库在洪峰期能拦蓄多余的水量，而在枯水期需要水的时候又能平缓地把水放出。

顺便提一下，蓄水除为了今后的用水外，蓄洪还可减少水库下游的洪水灾害。因此，水库可以用于防洪或者还具有别的用途。前一种情况称为“防洪水库”；而后一种情况就称为“综合利用水库”。

## 2 防洪水库

防洪水库一般称之为减洪水库，以拦蓄一部分洪水流量的方式降低洪峰，保护下游地区，为了达到这一目的，流进水库的水是全部排放，直到排水流量达到下游河槽的安全泄量为止。超过这个泄量的入流量就蓄存在水库里，待洪峰消退后，再逐渐排放，为迎战下一次洪水恢复蓄洪容积。

如图 8.1 所示，水库下游的洪峰流量减小了一个 AB 值。防洪水库和蓄洪水库不同之处仅仅在于它需要一个大的泄洪能力，以便在洪水之前或之后可以迅速向下游放水。

防洪水库的类型。防洪水库有两种基本类型：①蓄水水库或拦洪区；②滞洪区或滞洪水库。

在溢洪道或泄水口处设有闸门和阀门的水库称为蓄水水库，反之，具有固定的敞开式泄水口的水库称之为滞洪水库。

滞洪区的功能和优点：

滞洪区通常具有一个无控制的溢洪道和一个无控制的孔口式泄水道，根据入流量的大小自动调节水库的出流量。这种水库的最大泄水能力等于下游河槽的最大安全输水能力。洪水发生时，水库就蓄满水并经泄水道放水。水库水位上升时，出流量随之增加。当洪水消退、进水量等于或小于出水量时，水位就不再继续上升。此后，水便自动地由水库泄出，直到放完蓄水为止。与有闸门控制的拦洪区相比，滞洪区有以下的优点：①节约了按装闸门的费用；②没有闸门，因而不存在人为事故和玩忽职守的可能性；③这种水库并不总是蓄满着水，水库最高水位以下的大片土地只是被暂时或偶尔淹没，尽管不允许居民在此落户，但是仍能有效地进行耕作。

蓄水水库的功能和优点：

蓄水水库的溢洪道和泄水道有闸门控制时，运用起来就具有较大的灵活性，能更好地控制洪水，提高这种水库的效益。因而，这类水库通常用在需要良好控制的大河流上。而滞洪水库则用于小的河流上。蓄水库能更好地控制和适当地调节下游洪峰，使它们不会同时叠加。这就是这类水库的最大优点，尽管这种水库的缺点是投资较大，闸门的安装和操作包含着人为误差的风险。

### 3　配水水库

配水水库是小型蓄水库，建在城市供水系统内。这样的水库按一定的流量用抽水的方式注满，在需水量最大的时期（称为临界需水期），它的供水流量甚至大于水库的入流量。因此，这样的水库可以使水泵或水处理厂以均匀的流量进行工作，在不用水或很少用水的时期，配水水库就蓄存水量；在最大需水临界期，它就从“仓库”里供水。

### 4　综合利用水库

设计和建造的水库，若不仅为一种目标服务，而且同时为多种目标服务，就称之为综合利用水库。按单一目标设计，偶尔也为其他目标服务的水库，不能称为综合利用水库；在设计时除考虑主要目标之外，也考虑为其他目标服务的水库，才能称为综合利用水库。因此，若设计的水库用于保护下游地区免遭洪水灾害，同时又兼顾蓄水，为供水、灌溉、工业用水和水力发电等提供多种目标服务，它就可以称为综合利用水库。

## 第九课　混凝土的特性

要讨论任何符合特定用途的建筑物所需要的质量，就必须研究混凝土的特性。要使混凝土的每一个特性都趋于完美的最可行的方法，在很多情况下都是不经济的。最理想的建筑物是：在该建筑物中，混凝土的设计恰如其分地强调了该混凝土的各项特性，不能孤立地追求个别的特性指标，比如最大可能的强度。

虽然，在设计中获得混凝土最大强度不是惟一的标准，但是测定混凝土立方体或圆柱体的抗压强度为保持均一的混凝土质量标准提供了一种方法，事实上，通常也是这么做的。因为任何特定配合比的混凝土的其他特性在一定程度上都与抗压强度有关，抗压强度作为单项控制试验仍然可能是最简便的和最有益的。

在预制构件中，测试已硬化的混凝土毫不困难，因为测试过程中人们可以挑选完整的构件，如有必要的话，还可破坏这一构件。可以从已竣工的建筑物的某些部位中采样试验，不过这样做费用昂贵，还有可能减弱建筑物的强度。所以，习惯上在浇灌混凝土时用新拌制的混凝土作成试件进行试验，以估计建筑物的混凝土的特性。这些试件可以按英国1970年颁布的第1881号规范进行捣实和养护。但要在这两方面，精确地模拟建筑物的施工条件是不可能的。由于抗压强度还受试件的尺寸和形状以及所在建筑部位的影响。因此立方体试件的抗压强度没有必要与同样体积的混凝土块的抗压强度一样。

### 1　抗压强度

混凝土的抗压强度可以达到约80牛/毫米$^2$（12000磅/英寸$^2$），主要取决于水和水泥的相对比例——水灰比，以及混凝土的密实程度。当混凝土的配合比（水泥：砂：粗骨料）大致为1：2：4时，在很好的现场养护条件下，28天的抗压强度在20～50牛/毫米$^2$之间。某些类型的预制混凝土构件，如铁路轨枕，28 天的抗压强度可达 40～65 牛/毫米 $^2$，因为这些构件是用水灰比较低的富拌合物浇筑的。

除了水灰比和密实度以外，还有许多因素影响着混凝土的抗压强度。较重要的因素是：

（1）水泥的种类及其质量。混凝土强度增长的速率和极限强度均受其影响。

（2）骨料的表面特征。一个重要的事实表明，因某些骨料拌和的混凝土，比采用光滑的河底砾石拌和的混凝土具有更大的抗压和抗拉强度。

（3）养护的好坏。过早的干燥能使混凝土的强度损失大约 40%，因此，在现场施工中和试验过程中，混凝土的养护都是非常重要的。出于这个原因，我们必须严格遵守英国 1881 号规范关于混凝土立方体试件的养护方法。

（4）温度。一般来说，混凝土硬化的速度随着温度的增加而增加，在冻结温度下，抗压强度在一段时期里会很低。

（5）龄期。在正常情况下，混凝土的强度随着龄期的增加而增加，而增加的速度取决于水泥的种类。例如，用高铝水泥拌制的混凝土 24 小时的抗压强度与用普通硅酸盐水泥拌制的混凝土 28 天的抗压强度相等。但是混凝土的硬化速度很慢，要持续好几年。

以上是指极限静荷载的情况。当混凝土承受重复荷载时，会在小于极限静荷载的条件下破坏，这就是疲劳效应。许多研究人员已经证实；在承受几百万次的周期性加荷以后，混凝土的抗压疲劳强度只有极限静荷载的 50%～60%。

## 2 抗拉和抗弯强度

混凝土的抗拉强度随时间而变化，在初期为抗压强度的 1/8，在后期大约为抗压强度的 1/20，在钢筋混凝土结构的设计中通常不考虑抗拉强度。然而，抗拉强度对抵抗由于水分和温度的变化而导致混凝土开裂起着非常重要的作用。混凝土路面和飞机场常常要做抗拉强度试验。

混凝土轴向抗拉强度的测定是很困难的，因而很少进行这种试验。但有两种更加实用的评估抗拉强度的方法。一种是给出弯曲时抗拉强度的量度，通常叫做抗弯强度。英国规范（BS1881：1970）中详细说明了有关混凝土抗弯试件的制作和养护方法以及试验方法。当骨料的最大粒径为 40 毫米时，试件的标准尺寸为 150 毫米×150 毫米×750 毫米（长），如果骨料的最大标称粒径为 20 毫米时，试件尺寸可用 100 毫米×100 毫米×500 毫米（长）。

通过放置在 1/3 跨度处的两个辊筒施加荷载，直到试件破坏为止。边缘纤维应力即顶部的压应力和底部的拉应力，可以用一般的梁的计算公式进行计算。由于梁的抗拉强度远比抗压强度小，所以梁在受拉条件下会发生明显的破坏。英国规范（BS1881：1970）给出了断裂模量的计算公式。梁的试件有时用在现场，可以快速测定断裂模量和抗弯强度。然后，再用已经断开的两个半截试件进行抗压试验，这样，除了抗弯强度外，还能近似地用同一个试件测出抗压强度。英国规范（BS1881：1970）介绍了这种试验方法。

断裂模量值用于素混凝土道路和飞机跑道的某些设计方法中，因为在路面和跑道上，集中荷载分布在较大的面积上，所以混凝土的抗弯强度是足够的。

最近刚提出一种试验方法，沿圆柱体的直径加压，使圆柱体劈裂，所得结果命名为劈裂抗拉强度。英国规范（BS1881：1970）对这种试验方法有详细说明。为了使载沿圆柱体全长均匀分布，试验机械装有一个附加的支杆。将 12 毫米宽、3 毫米厚的胶合板条插在圆柱体和试验机支承面之间。根据破坏时的最大外加荷载由下式计算劈裂抗拉强度。

$$f_t = \frac{2P}{\pi l d}$$

式中：$f_t$为劈裂抗拉强度，牛/毫米$^2$；$P$ 为最大外加荷载，牛；$l$ 为圆柱体的长度，毫米；$d$ 为直径，毫米。

就抗压强度而言，重复荷载减小了极限强度，所以，在重复荷载作用下受弯时的疲劳强度是静力极限强度的 50%～60%。

### 3 剪切强度

实际上，混凝土的剪切总是与因弯曲而引起的受压和受拉同时出现，即使在试验过程中也很难消除弯曲因素的影响。

## 第十课 钢筋混凝土的基本概念

虽然混凝土已广泛应用于房屋建筑、桥梁和其他许多工程结构物中，但它的力学性能却很不理想。例如，它的强度不是很高。建筑用混凝土的抗压强度通常是 20～40MPa（大约 3000～6000lbf/in$^2$）。这比建筑结构物中所用的大部分木材的抗压强度稍低些。混凝土的抗拉强度非常低，大约是抗压强度的 1/10，因此大多数结构构件不能用素混凝土制作。混凝土短期受压时的弹性模量相当高，为 20000～30000 MPa（大约是钢弹性模量的 1/10）；然而由于徐变和收缩，混凝土在长期中又会产生较大的变形，所以其有效刚度要低得多——可能只有瞬时刚度的 1/3 到 1/4。

混凝土在工程结构物中所以能得到广泛应用，是因为它比目前常用的其他建筑材料便宜。其抗拉强度低的缺点则用加配钢筋以形成称为钢筋混凝土的复合材料来克服。虽然钢筋不能防止混凝土受拉区裂缝的产生，但是确实能限制裂缝的开展，因此提供了抵抗内拉力的有效手段。所需钢筋的数量，与混凝土的体积相比往往是很小的，所以钢筋混凝土结构的总造价在市场上的竞争力仍很强。

钢筋的主要用途是承受内拉力，因此钢筋要放在梁内靠近受拉面，即在跨中正弯矩区靠近下面，而在内支座负弯矩区靠近上面，如图 10.1 所示。在钢筋混凝土结构设计中，要在所有可能出现裂缝的区域都放置钢筋，这一点很重要。因而在由于剪切和弯曲联合作用而使梁产生斜裂缝的区域要放置成矩形配置的垂直和水平钢筋。为了便于施工，纵向钢筋（主筋）和横向钢筋（箍筋）可以预先绑扎成钢筋骨架。

在楼板中，钢筋往往是沿两个主跨方向成直角配置，以抵抗每一方向由于受弯而产生的拉应力。为了便于施工，板筋常常采用焊接钢筋网。

虽然钢筋主要用来承受由外荷载产生的内拉力，但它还有别的用途。钢筋承受压力的能力比混凝土强得多，所以当构件的总尺寸受限制时，有时也用钢筋来加强受压区的抵抗能力。这时在所有受压构件中都放置纵向钢筋，这样的构件除承受轴向压力外往往还受弯，这时配置在构件每一面的纵向钢筋或是受拉，或是受压。横向钢筋用来在浇筑混凝土时保持纵向钢筋的位置，以后当钢筋承受压力时，则用来防止钢筋向外屈曲。同样，柱子的钢筋也可预先绑扎成骨架。

不仅是由于外荷载，而且也由于温差以及不均匀收缩或收缩受阻，混凝土都会开裂，因而要配置辅助钢筋以控制这样的裂缝，这些裂缝可能很难看，甚至会发生危险。

插图译文：图 10.1　梁内纵向主筋　（*a*）简支梁；（*b*）连续梁。

# 第十一课　坝

据可靠记载，世界上第一座坝是公元前 4000 年以前在尼罗河上修建的。它使尼罗河改道，并为古老的孟菲斯城提供城址。至今仍在使用的最古老的坝是 16 世纪修建的西班牙阿尔曼扎坝。随着岁月的流逝，各种建筑材料和施工方法得到了改善，修建努列克这样的大坝才成为可能。该坝正在前苏联境内靠近阿富汗边界的瓦赫什河上施工，是一座高达 1017 英尺（333 米）的土石坝。大坝失事可能造成生命财产的严重损失。因此，坝的设计和维修通常是在政府监督下进行的。美国有 3 万多座坝由各州政府控制着。1972 年（美国）联邦大坝安全法（PL92—367）规定，必须由合格的专家对大坝进行定期检查。在 1976 年 6 月爱达荷州提堂大坝失事后，美国对大坝安全更为关切。

## 1　坝的类型

坝按其形式和建筑材料分为：重力坝、拱坝、支墩坝和土坝。前三种坝通常是用混凝土浇筑的。重力坝依靠自重维持稳定，通常在平面上呈直线状，不过有时略带点弧形。拱坝通过拱的作用把水的水平推力中的大部分传给拱座，因此，它的横截面比重力坝单薄些。拱坝只用于崖壁能承受拱作用所产生的推力的狭谷中。各种支墩坝中最简单的是平板坝，它是由许多支墩间隔地支撑着倾斜的面板。土坝是一种由土或石料填筑而成并借助于不透水的心墙或上游铺盖防渗透的土堤。在一座大坝的结构中可包含不止一种坝型。弧形坝可以把重力作用和拱作用结合起来，以利坝的稳定。长坝常常有一个包括溢洪道、泄水闸在内的混凝土坝段，其余坝段是用土或石填筑的副坝。

对既定坝址选择最佳坝型是一个关系到工程可行性及其造价的问题。工程可行性受地形、地质及气候条件所支配。例如：由于混凝土遭受冻融作用的交替影响而引起剥落，因此在低温地区常避免采用断面单薄的混凝土拱坝或支墩坝。各类坝的造价主要取决于能否在工地附近取得建筑材料和各种运输工具能否进入。大坝有时分期建造，第二期或以后几期工程，往往在第一期以后需要 10 年或更长的时间。

坝高定义为路面或溢洪道顶与基坑最低点之间的高程差。不过，引用的坝高值常常是用另外一些方法确定的，往往取原河床以上的净高度作为坝高。

## 2　作用在坝上的力

坝必须是相对不透水的，并能经受得住作用在这上面的各种力。这些作用力中最重要的是重力（坝的重量）、静水压力、扬压力、冰压力及地震力。这些力传给坝基和坝座，而坝基和坝座则对坝体产生一个大小相等方向相反的基础反力。某些特殊情况下还要考虑水库中泥沙沉积对静水压力的影响以及坝顶溢流所产生的动力作用。

坝的自重是其体积和材料比重的乘积。该力的作用线通过横剖面的形心。静水压力可同时作用在坝的上游面和下游面。静水压力的水平分力 $H_h$ 是作用于坝面垂直投影上的力，对于单位宽度坝体而言其值为：

$$H_h = \frac{rh^2}{2} \tag{11.1}$$

式中：$r$ 是水的比重；$h$ 是水深。该力的作用线在坝基以上 $h/3$ 处。静水压力的垂直分力等于坝面正上方的水重，并通过该水体的重心。

处于压力作用下的水必然要在坝和坝基之间流动，因而产生了扬压力。扬压力的大小取决于基础的特性和施工方法。经常假定扬压力从上游面（坝踵）处的全部静水压力直线变化到下游面（坝趾）处的全部尾水压力。根据这一假设，扬压力 $U$ 为

$$U = r\frac{h_1 + h_2}{2}t \tag{11.2}$$

式中：$t$ 是坝基的宽度；$h_1$ 和 $h_2$ 分别是坝踵和坝趾处的水深。扬压力的作用线通过压力梯形的形心（图 11.1）。

一些坝的实测资料表明（图 11.2）：扬压力比公式 11.2 所给出的值小得多。对扬压力的分布有各种不同的假设，美国垦务局认为重力坝的扬压力呈直线变化，在坝踵处为全部扬压力的 2/3，到坝趾处为零。坝踵附近通常设有排水装置，以便排除渗流水量，减小扬压力。

# 第十二课 土 石 坝

## 1 坝型的选择

一般来说，土石坝有两种类型：土坝和堆石坝。坝型的选择取决于能从需要开挖的地点和可用的料场处取得合用材料的情况。应该注意，根据岩石的物理特性，堆石能逐渐变化为填土，因而不能对土石料作出严格而固定的分类。那些软弱和在开挖填筑时容易碎裂的岩石可归入填土类。而坚硬和不会大量碎裂的岩石，则列为堆石类。

一座土坝的选定和设计都有赖于设计人员的判断和经验，而且在很大程度上是属于经验性的。各种稳定和渗透分析方法，主要是作为证实工程师的判断而使用的。

## 2 超高

所有的土坝都必须有一个足够的额外高度，称为超高，以防止库水漫顶。超高的高度必须足以在波浪作用、风力壅高和地震影响下，不会导致坝的漫顶。除了超高外，对于坝建成时发生的坝体和地基沉陷，还必须在高度上留有余地。

## 3 坝顶宽度

土坝的坝顶宽度一般用常规设备便于施工的填筑宽度来控制。通常，坝顶宽度应不小于 30 英尺。如果存在着大规模塌方进入水库，或者有因地震使岩块倒落而引起波浪漫顶的危险，则需要采用抗冲刷的材料填筑更宽的坝顶宽度。

## 4 定线

土坝的坝轴线选定应尽量使建设费用降到最少，但是也不能因此引起坝体发生滑动或开裂。一般说来，一条横跨河谷的最短直线，可能满足要求。但是，当地的地形和地基条件可能要求采用另外的方案。位于狭谷的坝，常采用向上游拱出的坝轴线，以便在坝体受库水压力作用而发生变形时，能使坝体压紧，从而尽量减少其横向开裂。

## 5 两岸坝座

一般有三个问题与土坝坝座有关：①渗透；②不稳定；③坝体的横向开裂。如果坝座是

由透水的沉积土构成，就可能需要建造一道上游不透水铺盖和下游排水设施，以尽量减少和控制坝座内的渗透。

在坝座岸坡很陡的地方，特别在边坡突变或有陡壁处，那里的坝体填土会有产生横向裂缝的危险。这个问题可以用开挖坝座放缓边坡来处理，这样的处理在不透水区和过渡区特别需要。过渡区，尤其是在上游侧的过渡区，必须用粘着力很小或无粘着力且颗粒级配良好的土料来填筑，这种土料如发生横向裂缝时能自行愈合。

### 6 分期施工

土坝的分期施工往往是可能的，而且在一些情况下是必须的。要求这样施工程序的因素是：①河谷宽阔，可以允许导流或泄水工程与一部分坝体同时施工；②地基软弱，要求坝体不要过快填筑，以防止地基中产生过大的应力；③料场潮湿，要求放慢施工，以使土料能通过固结作用来增加抗剪强度。在某些情况下，可以需要增设基础排水设施或填筑排水砂井，或采用水平透水的排水铺盖。

### 7 坝体的土料

大多数土料适用于坝体填筑。然而，在物理和化学性质上也有一定的限制。含有过多盐分或可溶物质的土料，不可使用。在土料里，不应存在大量的有机质成分。褐煤若能通过填筑而充分分散，无自燃之虞，就不防碍使用。而具有高度流限的肥粘土，多半难以施工，必须避免应用。

### 8 压实的要求

不透水和半透水的土料的强度取决于压实的密度。压实密度又取决于土料的含水量和压实设备的重量。因此，料场土料的含水量和在堆筑前或堆筑后而未碾压前的填土实际含水量变化，都会影响坝体的设计。如果天然含水量太高，可以在料场用排水或将土料耙松的办法来减低。如果土料太干燥，则须在料场用洒水或泡水的办法把土料湿润，然后再让土料在使用以前保持稳定的含水量。填筑时的含水量范围一般介于比标准普氏最优含水量低2%到高2%～3%之间。透水性土料至少应压实到相对密度的80%。

如果需要，应该变换填筑层的含水量、铺层厚度、碾压遍数和碾压机的型式等，进行填筑试验。对于坡度很陡的坝座部位，必须用薄层填筑，并用手扶打夯机夯实。所有外悬突出部位，均应在填土堆筑以前挖除或用贫混凝土填平。

### 9 观测仪器的类型

观测仪器的类型取决于工程的规模和复杂性。通用的装置是：①测压计；②表面位移标志；③沉陷量测仪；④测斜仪；⑤内部位移和应变指示仪；⑥压力盒（压力传感器或压应力计）；⑦地震加速度仪；⑧在管道接头和其他混凝土结构上的位移标志。

## 第十三课 岩基上的混凝土重力坝

设计人员在设计任何坝时，都必须对有关坝址的情况及其对结构物的影响作出一些基本假定。坝址的查勘为工程师拟定这些假定提供了许多资料。这些假定是安全设计的基础。有关小坝设计的一些主要假定包括：扬压力，渗流控制措施，河槽刷深以及下游坝趾的冲蚀，地基条件和施工质量。其他一些附加假定应包括泥沙荷载、冰压力、地震加速度和波浪力。

### 1 安全系数

安全系数应根据经济情况来考虑。安全系数大，导致结构费用昂贵；而安全系数小则可能引起失事，也会导致很大经济损失。只有在适当地确定坝内和作用在坝的滑动、倾覆以及应力超限的作用力之后，才能得到合适的安全系数。

## 2 倾覆

通常，抗倾覆安全系数在2～3之间，较小的坝常取较大值。如果算出的安全系数小于2，应修改坝的断面以提高安全度。重力坝很少因倾覆而破坏。因为任何一种倾覆的趋势，都更可能引起滑动而导致破坏。抗倾覆安全系数是绕坝趾的复位力矩和倾覆力矩的比值，可用公式表示为：

$$FS_0 = \frac{W_c \times l_1 + W_w \times l_2}{P \times l_3 + U \times l_4}$$

式中：$W_c$ 为混凝土的重量；$W_w$ 为斜面上的水重；$P$ 为水对坝向下游的推力；$U$ 为上托力；$l$ 为各力的力臂。

此外，如果上游面的扬压力超过任一水平断面上不计扬压力算得的垂直应力，则扬压力将使绕坝所假定的水平断面下游点倾覆趋势大大增加。在这种情况下，如果引起的拉应力小于混凝土内和地基材料内的允许应力，则该坝仍可认为是安全的。这个假定是基于施工工艺良好和结构内所有水平面上都具有抗拉强度作出的。

## 3 滑动

工程师们衡量向下游移动的安全性有三种方法。每一种都有优点，并且所考虑作用力之间的关系大体上都相同。尽管这几种方法算得的结果都安全，但它们有很大差别。这三种方法是：①滑动安全系数；②安全系数；③剪摩安全系数。必须将这三种方法区别清楚。每种方法的基本目的都是要求得一个安全系数，当越出此系数范围时，坝就有向下游滑动的危险。

重力坝具有水平地基面时，其滑动系数等于基面垂线和地基反力合力之间夹角的正切。在小坝上，滑动系数的计算可取水平力总和$\sum P$和垂直力总和$\sum W$（包括浮托力$U$）的比值，或

$$f = \tan\theta = \frac{\sum P}{\sum W - U}$$

如果这种方法求得的$f$值等于或小于静摩擦系数$f'$，则认为坝是安全的。计算中假定取单位宽度为1英尺。各种地基材料的滑动系数安全值列于表13.1中。

表13.1 不同地基情况的允许滑动系数

| 材 料 | 滑动安全系数 | 建议最小安全系数 | 剪摩系数 |
|---|---|---|---|
| 混凝土与混凝土 | 0.65～0.8 | 1～1.5 | 4 |
| 混凝土与坚硬岩石（表面清洁、不规则） | 0.8 | 1～1.5 | 4 |
| 混凝土与岩石（有些层理） | 0.7 | 1～1.5 | 4 |
| 混凝土与砾石、粗砂 | 0.4 | 2.5 | — |
| 混凝土与砂 | 0.3 | 2.5 | — |
| 混凝土与页岩 | 0.3 | 2.5 | — |
| 混凝土与粉砂、粘土 | * | 2.5* | — |

* 安全系数需要通过试验确定。

抗滑安全系数$f_s$的定义是静摩擦系数$f'$和地基面垂线与地基反力夹角的正切之间的比值，即

$$f_s = \frac{f'}{\tan\theta} = \frac{f'(\sum W - U)}{\sum P}$$

这个方法也假定剪力为附加安全度。对于岩基上断面偏保守的重力坝，抗滑安全系数一般在1～1.5之间。计算中计入扬压力和地震力后，安全系数可降低到接近1。这些数值都是指水平面上的抗滑安全系数。如果地基倾向下游，则基面上的抗滑安全系数会相应降低。设计人员常采取在槽或岩基中浇筑混凝土的办法来减少坝滑动的趋势。

另一个许多工程师乐于采用的方法，是在安全系数中包括剪力的计算。剪摩关系式为：

$$SSF = \frac{f'(\sum W - U) + b\sigma}{\sum P}$$

式中：$b$ 为所考虑剪切面上的底宽； $\sigma$ 为剪切面上的一种或几种材料的允许工作剪应力。

这种方法算得的安全系数应该接近正常结构计算中使用的数值。对于混凝土和岩石或混凝土之间的滑动，静摩擦系数常假定为0.65～0.75。混凝土的工作剪切力 $\sigma$ 和混凝土的抗压强度有关。混凝土的单位剪切强度约为标准圆柱体压缩破坏应力的1/5。对坝体混凝土来说，即为400～800磅/英寸 $^2$。如果计算中取工作应力为100～200磅/英寸 $^2$，则安全系数还有4。除非小坝的混凝土确实事先做过试验，否则不宜采用更大的工作应力。

## 4 混凝土的内部应力

混凝土和地基材料内的应力必须保持在规定的最大值之内，以免引起失事。在正常情况下，如果采用适当的混凝土配合比，则小坝混凝土内引起的应力都小于实有的强度。一种能保证耐久性的混凝土配合比，通常都具有充分的强度，要以对于抵抗应力超限提供足够的安全系数。

地基材料的应力超限问题也必须加以研究。这对于在有节理的岩石和软基（例如砂和砾石）上的小坝是确实需要的。设计人员应核对当地有关允许承载压力的规范，并且和有水平的工程师商量来鉴定地基材料。表13.2中列有设计混凝土小坝时，供初步研究和参考用的允许承载值。

**表13.2　各种地基材料的加权渗水比和承载值**

| 材料 | 朗氏加权渗水比 | 巴氏系数 | 允许承载值（吨/英尺 $^2$） |
|---|---|---|---|
| 极细砂或粉砂 | 8.5 | 18 | 3（密实） |
| 细砂 | 7.0 | 15 | 1（松散） |
| 中砂 | 6.0 | — | 3 |
| 粗砂 | 5.0 | 12 | 3 |
| 细砾 | 4.0 | — | 5 |
| 中砾 | 3.5 | — | 5 |
| 砾石和砂 | 3.0 | 9 | 5～10 |
| 粗砾（包括卵石） | 3.0 | — | 5～10 |
| 巨砾（带有一些卵石、砾石） | 2.5 | — | 10 |
| 巨砾、砾石和砂 | — | 4.6 | 5 |
| 软粘土 | 3.0 | — | 1 |
| 中等粘土 | 2.0 | — | 4 |
| 硬粘土 | 1.8 | — | 6 |
| 极硬粘土或硬盘土 | 1.6 | — | 10 |
| 良好岩石 | — | — | 100 |
| 成层岩石 | — | — | 35 |

# 第十四课　拱　　坝

拱坝最适合于修建在岩石峡谷中，它是一种控制河道水流经济而有效的建筑物。一座拱坝的承载能力足以使设计人员用较少的材料而仍能建成极为安全的结构。

## 1　理论

拱坝的一般设计理论比较新颖，同时在获得更多的资料之后，理论的变化也很迅速。工程师们曾慎重地应用数学理论、力学定律和弹性理论，以减小拱坝的厚度，并取得了很大的经济效果。

从历史发展看，拱坝分析方法从圆筒理论开始，发展到多个独立拱圈分析、拱冠悬臂梁分析，以及多悬臂梁分析。

拱冠悬臂梁和拱作用的理论已被证明能满足拱坝分析用。在这一分析方法中，水平水荷载在拱环和垂直悬臂之间分配，坝内拱和悬臂在交会点上的挠度可以通过改变坝的形状来调整，以使二者的弹性挠度近乎相等。当两个挠度近乎相等时，所算得的应力即作为坝内的实际应力。坝的形状或形式会改变荷载的分布。由于坝的形状通过试算法修改确定，所以这个方法就称作试荷载分析法。

正在研究的另一种方法是采用壳体理论来作拱坝分析。用于拱坝分析的大多数联立方程式，除了那些不规则边界条件外，都可以很快列出。联立方程式的求解可用电子计算机。

温度变化对柔性拱影响可能不大，但在厚而平坦的拱上却很重要。再则温度和收缩效应二者还可能在拱的整个厚度上是变化的。

这就很合乎逻辑地要在设计和施工技术中设法使温度应力减至最小。设计人员在选择拱形式时，要考虑到温度变化，以使内部拉应力减至最小。

拱坝设计中很少考虑倾覆和滑动问题。设计人员主要关心的是剪切、压缩和拉伸应力。现在已有详细的分析方法和计算机程序。工程师团、垦务局和几家咨询公司都备有这种程序。

## 2　基本导则

设计人员可以借助于以往的拱坝设计经验来进行设计。在初步选择拱坝时，需要一定的技巧来布置一些结构物及其附属设施，以取得经济效果，同时不危及坝的基本安全。

对于坝顶弦长为坝高的2～3倍的峡谷，拱坝是经济的。河谷较宽则需要增加拱的厚度，以保持适当的应力。即使长高比远远超过5，拱坝在经济上和结构上仍能胜过其他型式的坝。

坝顶拱圈的厚度通常取决于耐久、交通要求或冰压力。一般规定在任何高程处的拱厚度要等于或大于0.02$R$，$R$为拱的中心线半径。为了防止压屈，需要规定一个最小的厚度。

尽可能增大中心角就能获得经济效果。理论上，大约133°的中心角最有利。拱圈和岸坡拱座接触处的角不应小于30°。大于45°的角也允许，但拱的承载作用将显著减小，从而增加悬臂的作用，这又要求有较厚的拱，以使应力保持在允许限度之内。

为安全计，对于要求90天龄期强度为3000～4000磅/英寸$^2$的混凝土，设计人员可提高其设计允许应力为1000磅/英寸$^2$（抗压）和150磅/英寸$^2$（抗拉）。通过更好地了解结构性能，可知拱坝的安全度并不仅仅等于圆柱体强度除以设计允许应力。在模型上所作的可靠试验已经证明了分析结果，即安全系数大多都在圆柱体强度与允许应力比值的2倍以上。这样，实际上安全系数可达9～14。当设计人员对于拱的作用和拱座反力获得更多的了解之后，这个安全系数将有可能降低。

### 3 地基问题及其处理

拱座是拱坝的基础，必须承受拱传来的荷载——剪力、推力和弯矩。剪力不会很大，推力随高度变化，通常在坝高的中部最大。它可以通过断面上的应力的总和求得。拱座上的弯矩导致应力的不均匀分布。这就可能造成局部应力超过允许应力。拱座处的允许应力，是指岩石或混凝土的最大许用应力（取其小者）。对于拱座处过大的应力，可以设法在应力超过高程处修改拱的形状，或者在拱座处加大拱的厚度来降低。也可以采用扩大支承面积来降低应力，并且可以在拱内侧加贴角来做到这一点。只要有可能，修改形状比较可取，因为这样做对混凝土量影响最小。

如河谷在坝顶附近突然扩大，则可能需要设置推力墩将拱的推力传到地基。

基岩的弹性性质会引起一定的塑性变形，使结构内的应力重新分布。在拱冠梁分析方法中考虑了岩石的弹性。

### 4 结构模型

结构模型试验是拱坝的一种可靠的设计方法。如果运用得当，它可以预测坝内的应力。这种应力能和建成后的原型观测应力相符。

模型试验可以考虑坝的不对称性，采用周边或自由接缝和径向接缝，以及考虑地基和拱座的不规则情况等。试验得到的资料促使数学分析方法得到改进。再者，通过编制程序运用电子计算机来缩短数学分析所耗用的时间，实际上也就是提供一些数学模型。

## 第十五课 溢 洪 道

溢洪道相当于坝上的一个安全阀，它必须设计成能够泄放最大的流量，而同时能保持水库的水位在预定的水位以下。一个安全的溢洪道是极为重要的，许多坝的失事都是由溢洪道设计不当或者溢洪道容量不足造成的。溢洪道的尺寸和使用频率取决于流域的径流特性和工程的性质。入库设计洪水的选择和确定，则必须在充分研究流域水文因素的基础上进行。对于过坝水流的调泄，需要有合理审慎的设计，以避免生命财产的损失。

由于篇幅限制，这里不允许对洪水流量的水文分析作充分的讨论。但是提供了作工程初步研究用的最大流量估算资料。至于在工程的正式规划中对河川水流作年或多年利用时，则需要进行更详细的水文分析。

### 1 河川流量

对江河流量的研究包括：①确定经若干年一段时期内所能获得的水量；②确定溢洪道设计和大坝安全所必须宣泄的最大水量。

第一方面，要联系工程开发中的用水情况，按枯水期和丰水期来研究流量。优先的用水权必须调查清楚，并纳入研究中。要作出若干年一段时期内河川径流的累积曲线，以便确定可利用的水量。累积曲线表示一定时期内通过河道上某一地点的累积总水量。遗憾的是，大部分小河流上都没有足够的记录可用以编制水文资料。工程技术人员通常都从相邻的河流数据和雨量资料编制出综合曲线。从各种教材、杂志和报告中也可以得到各种估算河川流量的方法。

第二方面包括最大洪水流量的估算，用以确定所需要的溢洪道容量和坝的安全度。研究表明，洪水流量与不同时期中洪水的出现频率有关。这样，工程技术人员就能切实地估计洪水超过估算的设计流量所造成损失的危害性。

如果坝的失事会造成生命伤亡，则溢洪道必须有充分的容量，以防止最大可能洪水通过水库时发生失事。这种情况对于发生洪水时可能漫顶的土坝和堆石坝，尤为重要。混凝土坝如果在结构分析中遵守通用的安全系数，通常都能够经受一定程度的漫顶而不致失事。

对于不致危及生命的失事情况，如果有关部门充分估计到所冒的风险以及随之而造成的损失，则也是允许的。这种情况可能存在于小水库的低坝上。

用查图法可以快速估算出最大可能流量。由这些曲线定出的流量，应采用与该地区有关的水文资料加以修正。这些曲线系根据未经整治的河流上的非常洪水流量记录作出的。克里格（Creager）给出包络线的一般公式为：

$$Q=46CA^{(0.894A-0.48)}$$

或

$$q=46CA^{(0.894A-0.48)-1}$$

式中：$Q$ 为估算的最大洪峰流量，英尺$^3$/秒；$q$ 为相应洪水流量，以每平方英里流域面积上立方英尺每秒计；$A$ 为流域面积，以平方英里计；$C$ 为流域面积特性系数。

工程技术人员首先要取得列有全部近期洪水过程的资料，以及与研究地区有关的资料，然后才能采用从这些经验曲线上得出的洪峰流量。在初设布置中取系数 $C$=100，通常可以得到偏于安全的洪水可能值。

估算最大洪水的一种先进方法是将地区内能产生大洪水的暴雨转换成流域的洪水，将所得出的洪水加以分析，就可以用来确定洪峰流量和流量过程线。流量过程线是表示洪水产生特性的流量—时间关系曲线。如果用综合流域内的其他洪水形成特性（包括融雪）来研究最大可能降水量，则也可以得到相似的方法求出洪水过程线。

对于失事不会导致人员伤亡的建筑物，也可以采用比最大洪水小的洪水值。在蓄水量不多的较小结构上，容许在工程使用年限内发生预期的失事时，入库设计洪水可以采用 50 年或 100 年一遇的频率。

## 2 溢洪道型式

坝址条件对于溢洪道的位置、型式和组成部分有很大影响。而溢洪道的型式和对泄洪的各种要求对坝的构造型式也有影响。

溢洪道有六种常用类型：①顶部溢流式；②陡槽式；③侧槽式；④竖井或喇叭口式；⑤虹吸式；⑥闸门式。设计人员可以采用一种或几种型式的组合以满足工程的要求。

有些设计中，采用一种型式溢洪道供正常运行用，应付 50 年或 100 年一遇暴雨的洪峰。另一个非常溢洪道，则在发生正常设计假定中没有考虑的非常情况时，提供附加安全度。这类情况可以发生在洪水超过某一水位、溢洪道闸门发生故障或泄水道强制关闭时。非常溢洪道能防止坝的重要部位漫顶，对于土坝和堆石坝就显得特别需要。

顶部溢流式溢洪道很适合于混凝土坝，常用在那些坝顶长度可以满足所需泄洪容量的地方和坝基坚固或能抗冲刷的地方。有些坝上采用自由溢流式，另一些坝上则设置陡槽将水泄入下游河道。

陡槽式溢洪道常用在土坝或下游地基材料不良的地方。侧槽和竖井溢洪道则多用于空间受到限制的峡谷中。虹吸式溢洪道的水流特性适用于坝顶长度有限或要保持固定库水位的地方。当希望在大流量时减少坝的阻水影响和防止淹没过多时，都采用有闸门的溢洪道。

溢洪道可以是坝的一部分，也可以是另外一个单独建筑物。它的功能必须和坝结合成整体来考虑。坝的位置、尺寸和其他部分都影响溢洪道的位置和布置。最终的布置则要由溢洪道能满足水力条件和总的经济效益来决定。

# 第十六课 设 计 作 用 力

坝的首要功能是抬高水位，因此，坝所承受的主要外力是蓄水的压力。然而，也还有其他一些力作用在结构上。这些力都将在以下各节中加以讨论，计有：①水压力（外部的和内部的）；②泥沙压力；③冰压力；④地震力。

在重力坝上，坝的垂直重量是抵御水压力的主要作用力。在斜面支墩坝上，有一部分水荷载对结构起稳定作用。在拱坝上，水荷载通过拱的作用传到地基，重量则变为一个次要的稳定因素。主要作用力，其数值都能确切计算，在任何设计中必须加以考虑。设计的坝型不同，力的传递和安全系数也随之改变。

坝必须十分稳定以防止倾覆、滑动、应力超限和任何可能引起地基滑动的次生冲蚀作用。设计人员应仔细地考虑各种设计力，据以确定所需要的坝型和将要作用在结构上的各种力。

## 1 水压力

水的压强与水深成正比增加。正向作用于坝面的水压力可以用三角形的荷载分布图来表示。其合力作用于自水面到计算断面的 2/3 距离处，水压强的计算公式为：

$$p=wh$$

式中：$w$ 为水的容重（常用 62.5 磅/英尺$^3$表示）；$h$ 为自水面到计算点的距离，以英尺计。

总水压力由下式给出：

$$P_w=wh^2/2$$

小型重力坝的上游面一般是垂直的。因此水压力可用这一公式计算。当坝高增加时，在设计中常使垂直面略为倾斜。在这种断面上的全部垂直水荷载，相当于该断面上倾斜面上部三角形面积的水体重，其合力则通过该面积的形心。在小坝上，这个起稳定作用的荷载常常忽略不计。

在支墩坝设计中，坝面与垂直线呈一定角度的倾斜，水荷载对它起稳定作用。在这种情况下，水荷载传递到地面，荷载的合力增量，必须按不同的深度分段计算。

上托力以压力形式出现在坝和坝基的孔隙、裂缝和夹层中。在混凝土和基础材料内的孔隙中都充满了水，水在各个方向都作用有压力。扬压力的强度取决于水头，即水库的深度和从上游面到所考虑之点的距离。扬压力不仅发生在透水软基内，也发生在混凝土和岩石地基内。设计中采用的总上托力值，在很大程度上要根据地基特性、消除渗漏的措施、地基排水失效的可能性以及施工方法来判断。

对于宽缝坝和支墩坝，支墩之间的空间可以消除扬压力。然而，当这些坝建于透水软基时，必须注意防止地基材料通过排水孔道发生管涌。

在透水地基上，混凝土坝底部的扬压力和通过透水材料的渗流有关。经过透水材料的水流被摩阻力所阻缓。当坝建造在透水地基上时，必须考虑坝下渗透水流的强度和数量。扬压力对于透水地基上的所有坝都是重要的。

对于所有各种型式的地基，所用的降低上托力的方法都相同。这些方法包括：在坝的上游面灌注一道近乎不透水的帷幕；在靠近坝的上游面设置排水，以使水能自由流出；设置各种截水墙或者采用上述几种安全措施的组合。

在岩基中存在夹层、裂隙以及在坝透水地基中存在渗流时，都需要对上托力作出某些假定。对于岩基，为了安全，可假定从上游水压力直线变化到下游水压力，按此计算扬压力。扬压力作用在整个坝底面上。扬压力图形的其他任何变动都应通过电拟法验证或对现有类似建筑物作对比分析后采用。对于在透水地基上的坝，扬压力的具体图形，必须由流网分析决定。在流网分析中可包括适当设置的护坦、截水墙、排水系统以及其他控制扬压力的措施等。

任一点 A 的扬压力可由魏斯特加德（Westergaard）公式计算，即：

$$P_u=H_2+kx(H_1-H_2)/L$$

式中各项符号意义可参见图 16.1。该式中 $P_u$ 为水柱高度，以英尺计，转化成压强可乘以水容重 $w$。扬压力系数 $k$ 与排水系统的位置及其降低扬压力的效能有关。排水管设在靠近上游面灌浆帷幕之后，可以使 $k$ 值从 1.0（无排水系统）降到 0.5。降低后的压力，沿整个作用面变化，并假定呈直线降低，如图 16.2 中虚线所示。

在小坝上，采用地基排水措施往往是不经济的。在设计中都假定受全部扬压力作用。然而对于中等高度的坝，设计人员应考虑设置一个检查廊道，并装设减压排水管通入混凝土内和地基内。坝内排水廊道中的减压井都垂直布置，中心距约为 10 英尺。在地基内，减压井都从廊道中向下钻到 0.4～0.6 静水头或 2/3 截水墙或帷幕的深度。

# 第十七课　入渗的重要性及其过程

## 1　入渗的重要性

入渗是水进入土壤的运动过程。处于某确定状态下的土壤对水的吸收存在一个最大的吸收速率，其上限称为土壤的入渗能力。如果降雨强度低于这个能力，入渗速率等于降雨速率；如果降雨强度超过土壤吸收水的能力，入渗速率就等于土壤的最大吸水速率。超过入渗能力的降雨过剩部分会在土壤的表面汇集，并经地表面而流到河里。入渗速率和降雨强度一样，以每单位时间单位深度来表示。它们代表水在某一定时间间隔内浸入土壤的深度。

土壤的表面相当于一个滤池，它决定了雨水流到河道的途径。没有入渗的水会很快地流过地面，而进入土壤的水则在地下流得慢得多。因此，在确定暴雨径流量、暴雨径流发生的时间及其高峰流量时，土壤起着主要的作用。这些数据对于设计涵洞、桥梁和其他小型水工建筑物的水利学家来说是很重要的。在流经地表的过程中，水会侵蚀表土和地面上的一些重要的有机废料。水土保持学家所从事的工作，就是设法使这些地表径流渗入地下，或通过导流使它们安全地离开田野或农场建筑物。地形学者也从事于有关降雨强度的大小、频率及其入渗的立体分布特性的研究工作，因为地表径流是地貌变动的重要因素。

入渗到土壤的水在一定程度上控制着可供土壤蒸发蒸腾的水量。就植物生长而论，土壤含水量是由有效降雨量供给的。因此，生态学家和农学家根据对入渗与径流的研究来弄清楚植物要求与供水之间的关系。未经蒸发回到大气而入渗到地下的水进入地下水系统，并补给河流。入渗量的增加可以增大旱季时河水的流量，这对于供水、废水稀释和其他用途都是很重要的。

## 2　入渗过程

如果仔细考察一块土壤，或者在土壤断面挖坑，考察坑的各边，就会看到：土壤是由亿万颗砂子、淤泥和粘土的颗粒组成的，并且散布着许多大小不同的通道。这些通道包括收缩裂缝，虫孔，植物根的孔洞，土块或土壤团粒之间的空隙以及单个颗粒之间的微小空隙。这些裂缝、孔洞和微小空隙称为土壤孔隙。

当雨降到地表面时，部分或全部进入土壤孔隙。它由于重力和毛细管作用力而被吸入土壤。水因自由重力流而进入土壤的速率受孔隙直径的限制。当水沿这种孔隙流动时，它要受到水流阻力的作用，水流阻力随孔隙直径的减小而增加。在重力的影响下，水垂直向下经土壤纵向流动。另一方面，毛细管的吸力可以使水垂直上下流动或水平方向流动。毛细管的吸力能将水向上汲入窄的孔隙，就像在小玻璃管中将水向上汲的毛细管力比在大管中要大一样。虽然在孔隙很细的土壤中，这种毛细管力最强，但这些孔隙可能会细到使水流流过时的阻力相当大。在大孔隙中，例如在虫孔或植物根的孔洞中，毛细管力可以忽略不计，水则以自由重力流向下流动。当水向下流过这些通道时会受到侧向的毛细管力的作用，使水离开较大通道而进入较细的颗粒间的空隙中。因此，入渗包括三个相互依存的过程：经土壤表面进入，在土壤储存和经土壤而传递。

## 第十八课　腾　发　量

蒸发是水从液态变成气态的过程。蒸腾是植物将从土壤中吸收的水分以汽体的形式释放到大气中去的过程。到达地面的降雨有一半以上以蒸发与蒸腾相结合的形式——腾发，回到大气中去。在干旱地区，蒸发可以消耗掉大量的水库存水。

水面蒸发率与水表面水气压与上层空气大气压的差成比例（道尔顿定律）。在静止的空气中，气压差不久就变得较小，蒸发受水面水汽弥散速率的限制。由风引起的扰动和热对流将表层的水变为水汽，（从而）使蒸发得以持续下去。

蒸腾从本质上是指水分从植物叶表面的蒸发。因此，如果植物供水受限制，蒸腾速率大约等于自由水面的蒸发率。因此，估计的自由水面蒸发量可以假定代表有植被的土壤表面的潜在腾发量。

整个长时期的植物蒸腾总量主要受水的有效性限制。在雨量充足且年内分布均匀的地区，所有植物都将以大致相同的速率蒸腾，蒸腾总量的不同是由于各种植物生育期长短不同而引起的。对于供水受限制和季节性（降雨）的地方，植物根系的深度是很重要的。在这样的地区，当表土变干时，浅根系的草凋萎后便死亡，而深根的树木和作物将继续从深层土壤中吸取水分。深根植物在一年的蒸腾过程中将蒸腾掉大量的水分。在土壤含水量达到凋萎点之前，植物的蒸腾速率实质上不会因土壤水分的减少而减少。

腾发量有时被称为耗水量或总蒸发量，表示某一区域植物蒸腾、土壤蒸发、雪蒸发和自由水面蒸发的总水量。某一区域上的实际腾发量可以从本区总的供水量（包括降雨、地面和地下流入量以及引入的水量）减去从本区实测流出量（地表和地下流出量）来进行估算。当地表和地下储存量的变化显著时必须包括在腾发量之内。

用简单的方程表示腾发量与气象数据之间的关系，人们已经进行了许多偿试，如：

$$U_c=0.9+0.00015\Sigma(T_{\max}-32) \qquad (18.1)$$

式中：$U_c$ 表示耗水量，单位是英尺；$\Sigma(T_{\max}-32)$代表植物生长季节最高温度减去 32℉的累积值。当 $U_c$ 以厘米为单位，温度由℃表示时，方程（18.1）变为

$$U_c=27.4+0.00823\Sigma T_{\max} \qquad (18.2)$$

这些公式与多年某一阶段的年平均腾发量吻合良好。但由于蒸发过程的复杂性，很显然，我们不可能用简单的温度函数去很好地定义腾发量。

如前所述，某一区域的潜在腾发量可以用自由水面蒸发量来估算。实际腾发量等于有效水量受限时的潜在腾发量。对于一个有多种植物的自然流域，假设腾发速率确实随土壤水分的变化而变化是合理的。这是因为浅根植物将在深根植物前停蒸腾。我们可以用连续方程建立水量平衡方程：

$$P-R-G_0-E_{act}=\Delta M \tag{18.3}$$

式中：$P$ 表示降雨量；$R$ 表示地表径流量；$G_0$ 表示地下流出量；$E_{act}$ 表示实际腾发量；$\Delta M$ 表示水分储量的变化。

$E_{act}$ 可用下式估算：

$$E_{act} = E_{pot}\, M_{act} / M_{\max} \tag{18.4}$$

式中：$M_{act}$ 表示任何计算的土壤水分储水量；$M_{\max}$ 表示一假定的最大土壤含水量。这个典型的水量平衡方程可用于计算径流量和估算腾发量。

# 第十九课 灌 溉 方 法

灌溉方法可以分为四大类——地面灌溉、地下灌溉、喷灌和滴灌，和许多小类。地面灌溉是最古老的灌溉方法，现在还有大约 3/4 的灌溉土地仍然使用这种灌溉方法。地下灌溉的适用性是有限制的。喷灌可以用于任何气候条件，在湿润地区最受欢迎，而且仍在推广使用。滴灌是最新的灌溉方法，使水得到最有效的利用。

## 1 地面灌溉

地面灌溉包括沟灌和淹灌。沟灌用于行播作物，水在作物行间的小沟里流动，进行灌溉。用虹吸管或穿堤放水管向小沟供水，或用田间配水渠向一组灌水沟供水。

### 1.1 沟灌

沟灌会产生严重的冲刷，因为水在行间耕松了的、未加防护的土壤上流动。以升/秒为单位的最大不冲流量可用下式估算：

$$Q_{\max} = \frac{0.6}{i}\ (\mathrm{L/s})$$

这个公式只能用于沟坡大于 0.3%时，因为在坡度较缓的沟道里，流量通常受到沟道输水能力的限制，而不是受冲刷的限制。如果已经知道沟道土壤比一般土壤更易遭受冲刷，那么，式中 0.6 这个系数应该减小。

坡度大于 2%时，对于灌水沟在灌水时很难避免冲刷。流量太大会冲刷土壤；流量太小只能流动很短距离就全部渗入土壤了。等高沟克服了这个困难，因为这些灌水沟垂直于主坡方向布置，沟道坡度约为 0.5%。要格外细心以确保水流不会穿越等高沟间的田垄。在较浅的沟道里，如果流量过大就可能使沟道漫溢，并引起连锁反应，而产生冲沟。

灌水沟的长度受到在灌水时间内灌溉水流流动距离的限制。均匀的灌水要求灌水时间不长于总灌水时间的 1/4。例如：总灌水时间为 8 小时，灌溉水流应在 2 小时内到达沟尾；总灌水时间为 12 小时，灌溉水流应在 3 小时内流到沟尾。所以，灌水沟的最大长度取决于土壤的渗吸速度和抗冲能力、沟道坡度以及灌水定额（灌水深度）。表 19.1 列出了不同条件下适宜沟长的估计值。

按照前面几段简述的原理进行灌溉时，施用水的灌水效率可以达到 60%左右，其余 40%的水量损失于蒸发、灌水沟上游段和强透水土壤的深层渗漏以及沟道末端的废泄水量。灌溉水流接近沟道末端时，如果灌水员减小灌水流量，就能减少废泄水量。剩余的水量可以用到别的地方，多半用于需要短时间灌水的牧草。

前面介绍的沟灌原理也用于相近的灌水方法。称为垄沟的小沟用于不中耕的粮食作物和饲料作物。和行播作物相比，这种植被能更好地保护土壤。但是，垄沟太小，不能输送较大的流量。因此，垄沟的长度和沟灌法的沟长类似。

### 1.2 淹灌

淹灌有三个主要类型：格田淹灌、畦灌、漫灌。格田淹灌大概是最古老的方法，埃及 5000 多年前就使用这种方法。由于这种方法简便易行，现在仍然广泛使用，或使土地长期淹水，种植水稻；或使土地短期淹水，种植其他许多作物。

格田淹灌的田间工程是在每一个灌溉田块周围设置 15～50 厘米高的田埂，每块格田内部的地面高程应尽可能均匀相等——当然要限制在 5 厘米或 10 厘米的范围内。格田面积取决于地面高程变化、田块大小和耕作要求。田块大小要使灌入的水量能均匀地覆盖透水的土壤。格田的大小变化从为灌溉单棵树或蔬菜作物而设计的很小格田到种植水稻面积可达几公顷的格田。

格田的一侧必须设有足以淹没格田的供水渠道或其他供水设施，把水放入格田，达到预计的水深，然后，把流量减小到恰好能够保持稻田大约为 10 厘米的水深；或者完全停止放水，以利其他作物的生长。可以使格田里的水量完全渗入土壤，或对一些渗透性较差的土壤，在到达特定的入渗时间后，把多余的水排到较低的格田里。

畦灌可以描绘成长条形的格田，沿其长边方向具有微小的坡度。

灌溉水流从畦首进入，沿长边方向向下流动，畦田宛如一个很宽的灌水沟。畦宽在 3～30 米范围内变化，沿宽度方向必须接近水平，以便使全部面积都能得到均匀地灌溉。对相似的土壤和坡度，畦田长度和灌水沟的长度类同，见表 19.1。

适宜采用畦灌的地面坡度是：种植精耕作物时为 0.2%～2%；种植小粒谷物或饲料作物时为 4%～5%；种植牧草时约为 8%。采用畦灌的地形必须比采用沟灌的地形平坦，因此，常常需要在大面积范围内进行土地平整。把水放入几块畦田比放入许多灌水沟或垄沟花费的劳力要少，这就补偿了平整土地的费用。平缓的地形有利于穿过畦田进行横向收割。

畦灌的几种变通方式已经设想出来了。在坡地上进行灌溉时，田块可以整成有坡度的畦田，也可以整成水平的格田，这是一种范例。在另一种变通方式中，用渠道代替畦埂，堵塞渠道，放水入畦，进行淹灌。遗憾的是采用这种方法，渠道冲刷是一大难题，而且往往很难实现均匀灌水。

漫灌用于灌溉饲料作物，有时也用在未经平整的土地上灌溉小粒谷物。水沿渠道向下流动并被引出来灌土地，时常需要临时开挖小分水沟把聚集在洼地里的水再次分配。灌水员用铁锹挖沟、筑埂，把水引向仍是干旱的地方。

漫灌的用水效率和劳动效率都是很低的，它能用来灌溉那些不能采用其他地面灌水方法灌溉的土地。这些土地，可以是由于土层太薄或石块太多而不能进行土地平整，也可能由于不能充分证明进行土地平整的巨额投资是合理的。坡度达到大约 10%的起伏地形，并具有多年生的茂密植被时，也可以采用漫灌。

## 2 地下灌溉

地下灌溉（系统）可以看作是一个有控制的排水系统。通常使用明沟，有些工程也使用暗管。这些工程系统在雨季用于排水，在旱季用于补水，因此，地下水位总是处于一个被控深度。在粗砂土壤上种植浅根作物时，这个深度也许只有30厘米；在壤土地区，这个深度也可能达到120厘米。表层土壤应该是干燥的，而根系分布层的绝大部分土壤应该是潮湿的。这些农田可以在灌溉的同时进行耕作。

地下灌溉要求的条件是很严格的，因此，只有很少的土地能够采用地下灌溉。地面一定要相当平整，并且具有小于0.5%的坡度；底土必需是强透水土壤，但是它必须有一个浅水地层，在下面铺一个不透水层，以能维持一个较高的地下水位。土壤和灌溉水中的含盐量都必须很少，以防止土壤盐碱化。在冰川冲积平原、阶地或湿润、半湿润地区的三角洲地带常常具备地下灌溉所要求的条件。

## 3 喷灌

比起地面灌溉和地下灌溉，大型喷灌是相当新的技术，因为喷灌所需要的管道、水泵、动力机械都是近代才出现的。喷灌具有便于移动、对土壤和地形有很大的适应性、只需要很少或者根本不需要田间工程、能准确地控制灌溉水量等优点，所以受到普遍欢迎。灌水效率高，其结果是可以节省能源、避免冲刷，防止土壤中硝酸盐的流失。限制喷灌使用的缺点包括：设备费用和运行费用高，要在泥泞的土地上移动管道，如果灌溉水质不好，会对某些作物产生盐害，对某些作物会引起病害。

绝大部分的大田喷灌采用通用的旋转式喷头。虽然，在种植经济作物的农田里，可以把喷头安装在固定的位置上，但是，喷头通常是安装在移动的或可以移动的管道上。

## 4 滴灌

滴灌是最新的灌水方法，也是灌溉效率最高的方法，大约90%的灌溉水量可以被植物利用。它通过很细的塑料管直接向每一棵植物供水，这就是灌水效率高的原因。灌水是连续的或十分频繁的，因而，植物的根系总是生长在湿润的土壤中。

滴灌特别适用于灌溉树木或其他高大的植物。果园和葡萄园已经大量使用滴灌，各种行播作物也在使用滴灌，包括几种蔬菜和水果。植株间的土壤是干燥的，这是滴灌最大的优点。对于像草坪、牧草、小粒谷物这些茂密的植物，滴灌则没有优点。

一位名叫西姆查·布莱斯的以色列工程师在19世纪30年代就提出了滴灌这个想法，但是，直到塑料管道问世后，才出现了实用的滴灌系统。美国的滴灌面积由1960年的40公顷发展到1976年的50000多公顷。1976年全世界的滴灌面积约160000公顷。美国接近半数的滴灌面积在加利福尼亚州，其中一些滴灌面积上种植鳄梨，地面坡度为50%或60%。因为不产生径流，所以没有冲刷问题。

滴灌还有一个优点，就是与其他灌溉方法相比，它可以使用含盐量更高的水源，含盐量可以高达2500毫克/升。滴头供给的稳定流量把水中的盐分带到植物根系区域的外缘。在植株之间的干燥区域内，盐分含量变得很高，但是在根系区域内盐分并不增加。

滴灌节省水量，能使用含盐较多的水源，在除了质地极粗和极细的土壤以外的所有土壤上都能获得满意的灌水效果，可以在几乎所有的地形条件下进行灌水而不产生冲刷，而且不需要多少劳力。缺点主要是设备费用高，管道易被泥沙，盐分结晶或藻类植物堵塞。

滴灌系统一般包括一个控制箱，用以调节水压力、过滤灌溉水、添加化肥和除莠剂。也可以加入氯气以杀灭藻类植物。滴灌的水压力一般是 0.4～1 千克/厘米 $^2$，而喷灌要求的水压力一般为 1～8 千克/厘米 $^2$。滴灌系统上安装了一些控制设施，用以周期性增加压力和冲洗管道，以减少堵塞。

滴灌管道分为三级或四级，以提供许多所需的出水口。最末一级是直径为 12～32 毫米的柔性塑料毛管，或放在地面上，或埋在地面以下，通常管道上的小孔或通过滴头进行灌水。滴头使水流经过一段很长的螺旋形通道，减慢流速后，再从一个较大的滴水孔中流出，较大的滴水孔不容易被堵塞。

# 第二十课 灌 溉 渠 道

渠道定线和公路定线有许多相似之处，但是渠道定线更加困难，因为渠底坡度必须是下坡，而且要避免坡的频繁变化（因为坡度的变化会导致渠道断面的变化）。在一定的地形条件下，正确的渠道线路取决于允许的渠道坡度。渠道坡度过大会使流速过大，以至冲刷渠底和边坡。开始冲刷的流速决定于渠床材料和渠道横断面的形状。对细粒土壤的冲刷流速一般比对粗粒土壤的冲刷流速要小，但是也并非总是这样，因为当土壤中存在胶结物质时，就会大大提高它的抗冲能力。在使用过程中，渠床土壤逐渐压实，抗冲能力不断提高。夹带着磨蚀物质的水流对结构紧密的或加固了的渠床材料有更明显的冲刷能力。表 20.1 列出了各种渠床材料的最大允许流速的估计值。一种更完善的防冲方法是把边壁剪应力（单位面积上的拖曳力）和容许的单位拖曳力相对照。通过这种方法，可以得到一个平衡的渠道断面，即所确定的渠底宽度和渠道边坡使渠底和渠坡都不大可能发生冲刷。

如果渠道坡度太缓，流速可能很小，就会滋长水生植物，减小渠道的输水能力，水中悬浮的泥沙也会沉积。所以，如果地形允许，应使设计流速略小于最大允许值。

土渠的断面一般是梯形，边坡决定于渠堤材料的稳定性。确定渠道稳定边坡的方法和土坝坡的计算方法相同。表 20.2 列出了未用各种材料衬砌的渠道的标准边坡值。

考虑到渠道中泥沙淤积、水生植物减小渠道输水能力、风浪作用、渠堤沉陷、暴雨期间流量超过设计值等因素对渠道安全的影响，在渠道设计水位以上必须有一个渠堤超高，作为防护措施。主要通过挖、填平衡，达到节省挖方费用和土方工程费用的目的。可是，当运距很远时，借土或弃土可能是有利的。在山坡地带，为了实现挖、填平衡，渠道可能很深。图 20.1 给出了典型的渠道断面。

如果筑渠土壤透水性很强，渠道水位又很高时，衬砌渠道以减少渠道渗漏可能是经济的。无衬砌渠道的渗漏率主要受土壤性质和地下水位高低的影响。渗漏量可以用以下方法量测：①静水池法；②入、出水量平衡法；③渗流仪测定法。静水池法是用临时性的不透水隔墙隔离一个渠段，放水入内，水池内水的消失量减去蒸发损失量就是渗漏量。入、出水量平衡法是不可靠的，因为在入、出水量量测中的很小误差会在渗漏量计算中产生相当大的误差。渗流仪是用一根管子把一个倒置的、断面积为 2 平方英尺的金属杯和一个柔性水袋相连接，水袋开始时是装满水的，并淹没在渠水中，金属杯口朝下插入渠底。打开阀门让袋中的水不断流向金属杯，补充渗入渠底的水量，袋中水的损失量就是渗漏量。为了用这种方法确定渠道的平均渗漏量，必须在渠底的不同地点做许多次试验。

为了减少渠道的渗漏损失，可采用不同类型的衬砌，粘土、沥青、水泥砂浆以及钢筋混凝土材料已经得到有效使用。埋置塑料薄膜是一种有效而廉价的衬砌形式，施工时，先在渠道的边坡和底部喷洒沥青，然后在薄膜上覆盖厚约 6 英寸（15 厘米）的保护土层。水中含有的细粒泥沙也许能帮助渠道堵塞土壤孔隙。对于重要渠道，混凝土衬砌通常是令人满意的，因为它的耐久性很好。钢筋混凝土衬砌层的厚度为 2～8 英寸（5～20 厘米），具体的数值取决于渠道的大小及其重要性，标准钢筋含量在纵向为 0.5%，在横向为 0.2%，要按一定的间距设置防渗的施工缝。用于小渠道的砂浆衬砌常常是在钢制的网格内喷射砂浆或用活动模板进行施工。为了保证混凝土衬砌的成功，渠堤必须是稳定的，而且排水是良好的。在渠道放空时，衬砌层下面的扬压力会引起严重的破坏。衬砌得当的渠道的渗漏损失可以降低到 0.05 英尺/日（0.015 米/日）。渠道衬砌除了减少渗漏水量外，还允许有较大的水流速度和较小的渠道横断面，从而节约工程费用。

## 第二十一课　耕　地　排　水

耕地排水就是将某一地区多余的地表水排掉或将地下水位降至作物根区以下，以促进作物生长或减少土壤中盐分的积累。

耕地排水系统与城市暴雨排水系统有许多共同的特征。在城市不宜采用的明沟排水在农村排除地表水得到了广泛的应用，比暗管排水节省了很多投资。在适宜的条件下，明沟也可用于降低地下水位。然而，明沟间距太小会影响农事操作。因此，排除多余土壤水更为常用的方法是暗管排水。尽管在可能的地方用大的暗管代替明沟是一种趋势，但通常还是将暗管中的水排入明沟。这样就可将节省出的多余土地用于耕作，从而废弃了难看的且有时是危险的排水明沟。因为耕地排水通常是解决很平的或淤填土地上的排水问题，因此提供具有挡潮闸和抽水设备的排水枢纽对集水的最终排除是十分必要的。

耕地排水加速了水的径流，从而增加了排水区域下游的洪峰流量，因此在规划排水系统时应考虑这一结果的影响。湿地是重要的生物生存场所。湿地为迁移水鸟提供了栖息的场所。在沿海地区，湿地是许多重要的具有商业价值的水生物种的繁殖场所。对这样的耕地排水对环境所产生的影响需要进行认真的评价。

耕地排水与其他排水类型所要求的水文设计依据不同。耕地排水的目的是为了在一定的时间内排除一定体积的水。因此，为了控制盐分，（必须设置）地下排水（系统）来排除灌溉土地上多余的水分。众所周知，每个灌区都灌了一定体积的淋洗水，所以在两次灌溉期间排水暗管应能把这部分水排走。

排除多余降水（雨）的排水系统是按照在 24 小时内能排除的额定数量的水进行典型设计的。通常所说的排水模数或排水系数，其建议值大约是年平均降雨量的 1%。上面定义的排水模数通常是设计年最大 24 小时降雨的 1/4 到 1/2，也可按设计年 24 小时暴雨减去使土壤达到田间持水量所需的入渗量进行估计。如果灌溉土地的排水模数超过通过用洗浇水估计的定义数量，则应该采用较大的值。然而，通常最末级排水沟尺寸（4 英寸或 10 厘米）大于要求的输送设计流量的尺寸。

用大堤保护的土地由于河水入浸或河水从堤下入渗而常常遭受涝灾。在这种情况下，排水系统的设计流量应根据河段 1 年或 2 年重现期的渗漏量进行估计。如果河段高水位与雨季同时发生，则排水模数应加上渗漏量。

排水系统由干沟、分干沟和支沟组成。排水沟通常是不衬砌的。小型沟道可以用专用挖沟机进行施工，而较大沟道通常用拉铲挖土机挖掘施工，某些很大的沟道则用挖泥船进行施工。除非挖出的土用于筑堤，为排水沟提供加大流量的场所，否则，这些弃土应堆放在距沟边至少15英尺（4.5米）远的地方，这样它的重量才不致对沟岸的稳定产生影响。弃土堆减少了耕地面积，阻止了附近农田向排水沟中排水。如果可能的话，弃土应分散成薄薄的一层，但没有地方能够分散这些弃土。凡是在天然排水沟与人工排水沟交叉的地方，弃土堤（堆）都应留一个开口，而且沿排水沟方向，弃土堆至少每隔500英尺（150米）也要留一个开口。

排水沟的坡度小，其横断面尽可能接近最优水力断面，梯形断面是最常用的形式，边坡不陡于1∶1.5。砂质土要求边坡为1∶2或1∶3。有时在必须进行抽排的地方，排水沟应考虑采用非经济断面，以便为排水沟创造尽可能多的贮存空间，从而使抽水负荷降到最低。排水沟的坡度、位置和间距主要取决于当地的地形。排水沟的最小可行坡度大约是0.00005（3英寸每英里或5厘米每公里）。排水沟一般沿着自然坡度降低，但尽可能沿地界线布置。尽管在平地上可能要求沟的间距为400米，而支沟很少需要间距小于800米的。对有适当坡度的土地，其排水沟间距用1600米可能是适宜的。

排水沟的深度通过为6～12英尺（2～4米）。有末级排水沟的地方，支沟的深度必须足以使地下暗管里的水流入其中。类似地，分干和干沟也必须有足够的深度，以接纳比其小的排水沟的水流。如果地形是平的且排水沟相当的长，则要求排水干沟的深度可能太大，因此可将原排水系统分成两个或两个以上系统，以缩短排水沟的长度。对腐殖土或泥炭土，当水从土壤中排出时会产生相当大的下沉，因此排水沟必须按比例加深。

地面排水系统设计的基本步骤与城市暴雨排水系统没有多少不同。其设计步骤可以归纳如下：

（1）准备一张具有详细等高线的排水区地图，等高线的间距通常需要1英尺（0.3米）。

（2）选择排水系统出口的位置。如果有几个可能排水出口，就需要对这些方案进行经济分析。

（3）确定地下排水沟的排水模数，并估计排水明沟将要接纳的排水量。

（4）布置明沟排水系统（或排水干管系统），使之具有足够的尺寸，以便输送期望的流量。

（5）确定二级排水管的适当深度，并对末级排水管进行平面布置。常用的最小尺寸（4英寸或10厘米）的田间排水管一般具有足够的容量，只有干管和支管要求的尺寸才必须进行计算。

（6）地下排水管规划完以后，可能要求对干沟（干管）的第一试验方案进行审核。排水沟（管）采用最短可能路径法进行设计，从而使整个系统按费用最小原则进行规划。

（7）估计工程造价，并进行承包工程项目所必须的法律程序。

## 第二十二课 滴灌系统的应用方法

滴灌系统类型的选择和技术要求取决于水质、土壤类型、作物种类和气候条件等几个参数。在砂质土和用微咸水（灌溉）的情况下，滴头应紧靠作物布置。这样作物就能够在水分和养分可以利用的区域内建立一个良好的发达根系，而盐分的积累是最少的。然而，根系的体积将是小的，从而需要对作物进行频繁的灌溉。在每一灌水周期内的灌水总量（甚至农田每天不至灌一次）必须满足对有效根系提供适当的淋洗用水——每个滴头的流量至少是1.5～2升/小时。

滴头的间距一般取决于农田作物的种植密度。但还应考虑土壤类型和水质。例如，Arava地区西红柿延苗床方向两株间距为0.5米，因此两滴头的间距也是0.5米［图22.1（a)］，当辣椒用微咸水灌溉时，一垄辣椒一条滴灌带，支管放在5～10厘米深沟内，每个滴头旁种两棵辣椒。每一苗床安装2条或3条支管，滴头的间距是40～50厘米［图22.1（b)］。当用良好水质的水进行灌溉时，可在两垄辣椒之间只放一条滴灌带。辣椒和滴头之间的间距大约是25厘米，洋葱或胡萝卜的栽培要求整个苗床的土壤都是湿润的，要实现这一条件，每一苗床要安装两条滴灌带［图22.1（c)］。滴头的流量取决于土壤的类型，砂质土滴头流量为4升/小时，黄土滴头流量为2升/小时。

在干旱地区的果园，滴头要紧靠着果树安装，在Arava地区，为了给棕榈树提供适当数量的水，挨着每棵树安装1个滴头，每个滴头的流量是8升/小时，滴头间距大约是30厘米（沿3米长曲线布置），在芒果树种植园，在每1米长的滴灌带上安装5个滴头，每个滴头流量为8升/小时。在这些情况下，根系的生长取决于（采用不同的）灌溉制度所形成的淋洗区体积（的大小）。

在每年降雨超过150毫米的地区，无论果树的位置如何，滴头都要沿支管等间距布置，因为根系的生长是随冬季雨量的增加而增加，因此，要使根系生长就必须为其供水。

安装长管线更加经济。因为它允许用户减少灌水器的数量，简化田间滴灌系统的配置。但是，延长滴灌带的长度要求使用压力补偿式滴头，因为压力补偿式滴头延管线均匀供水。用具有长滴灌带的滴灌系统有利于大面积作物耕作，如棉花和玉米。

当使用长管线时，应特别关注以下参数：①与灌溉周期长度有关的系统要求的供水时间；②流向滴灌带末端的水流流速；③滴灌带的排水。

### 1 充水时间

对直径为16毫米或20毫米的长滴灌带，充满整个系统需要相当多的水量，并要求提供调节出水流量的压力。当灌溉系统开始充满时，经常发生水从管线始端的滴头流出，甚至在水到达管线末端的滴头前，有时流出了相当多数量的水。当两次灌水周期之间时间间隔较长，即超过3天，大约10分钟的充水时间是允许的；然而当灌水是在比较短的时间内完成，比如说1天或2天，充水时间大约可为灌溉周期的16%。为了减少系统要求的充水时间，应当安装具有大直径的阀和过滤器，以保证（通过）大体积的水流。

### 2 流速

滴灌带中的水从流速400～600厘米/秒变成水滴，所以滴头就有一个很高的被水挟带的微粒堵塞的危险。为了克服这一问题，一种特殊的设备——管道冲洗器已经被开发出来。这种冲洗设备安装在管线的末端，应用快速水流对管线具有成功清洗的可能。管道冲洗要连续进行直到系统中产生了稳定的水压力再关闭设备。

### 3 滴灌带排水

如果配水器安装在坡地的最低侧，在每次灌水结束时，农田较低处可能产生（土壤）过湿的问题。因此，在具有斜坡的地方，配水器必须安置在高地上。

# 第二十三课　水　轮　机

## 1　概述

人们可以利用三种基本方法从水中获得动力：即利用水的重力作用、水的压力作用或水的流速作用，以及其中任意两种或全部三种作用的组合。在现代实践中，培尔顿式水轮机或冲击式水轮机是惟一以单一的方法，即靠一股或数股高速射流的作用取得动力的一种水轮机。这种水轮机通常应用在高水头电站上。

水轮机的最高效率自从 1925 年前后达到 93%或更多一点以来，实际上已经没有再提高。就最大效率而言，水轮机已经达到了实际上所可能发展的极限。虽然如此，水轮机的实体尺寸和单机马力容量在近几年里已有了迅速而明显的增长。

此外，对于空蚀的起因和预防也作了大量的研究，从中取得的好处是能在高于以前认为合适的水头条件下，获得更高的比转速。这些进步（包括更大的机组，更高的比转速，以及设计上的简化和改进）的实际效果，已使水轮机赢得了长期作为最重要的原动机之一的重要地位。

## 2　水轮机的类型

水轮机可以归纳为两种基本类型：冲击式水轮机——它利用高速射流在任一瞬间仅冲击水轮机轮周的一小部分时所产生的动能；反击式水轮机——它利用充满在转轮和过水道中的水的压力和流速的联合作用得到动力。反击式系列又分成两种通用的型式：弗朗西斯式（有时称作反击式）以及旋桨式。旋桨式又进一步再分为定轮叶式水轮机和以卡普兰式代表的转叶式水轮机。

### 2.1　冲击式水轮机

在冲击式水轮机上，压力钢管中的水的势能转换成射流（从喷嘴孔口中射出）的动能。这种射流自由地射入水轮室内的大气中，并冲击在转轮的碗状戽斗上。每旋转一周戽斗进入射流、经过并从射流转出一次。在这段时间内戽斗承受着射流的全部冲击力。这种冲击力产生一个高速锤击冲打在戽斗上。与此同时，戽斗受到离心力的作用而有脱离它的座盘的趋势，由此而产生的应力以及水流在戽斗的碗状工作面上的冲刷作用都很大，因而需要选用能抵御水力磨损和疲劳的高质量材料，一般都采用青铜和韧化铸钢，只有水头很低时才能用铸铁。

### 2.2　弗朗西斯式转轮

就弗朗西斯式水轮机来说，来自蜗壳或水槽内的流速较低的水，通过位于转轮周围的导叶或一些闸门，然后流经转轮，并从转轮泄入安置在尾水位以下而不与大气相通的尾水管内。由于水充满所有的水道并作用在转轮的整个周围，因此，仅有一小部分动力来自水的流速所引起的动力作用，而大部分动力则都通过作用在转轮叶片前后工作面上的压力差取得。尾水管可以使能利用的水头得到充分的利用，这一方面是由于转轮下面垂直水柱所产生的吸出作用，另一方面是由于尾水管的出口面积大于紧接转轮下喉管的面积，从而使水流离开转轮叶片时的一部分动能得以利用。

### 2.3　旋桨式转轮

旋桨式机组最适用于低水头电站，在它适用的水头范围内，已产生了显著的经济效果。这种水轮机的转速比较高，以致使发电机的价格较低，并使发电厂房的水下结构和水上结构的尺寸都较小。低水头、小功率的旋桨式转轮，有时用铸铁来制造。水头高于 20 英尺时，都用一种更为可靠的材料——铸钢来制造。大直径的螺旋桨可用单个叶片固定在轮毂上制成。

**2.4 转叶式水轮机**

转叶旋桨式水转机是从定轮叶旋桨式水轮机发展而成的。卡普兰式水轮机是这类水轮机中为人们最为熟悉的一种。它的叶片可由液压伺服器调整到效率最大的角度。利用调速器上的凸轮能使叶片的角度随阀门的开启位置而变化，从而在所有各种满负载百分率情况下都能保持高效率。

由于转叶旋桨式水轮机组在闸门各种开度情况下效率都高，因此，它特别适用于那些必须在变负载和变水头条件下运行的低水头电站上。当然，这种机组的投资费用和维护费用要高于只能在一个最大效率点上运行的定轮叶旋桨式水轮机组。

# 第二十四课 水 力 发 电

法拉第曾经指出：线圈在磁场中旋转，就产生了电。因此，为了获得电能，我们必须产生使“线圈”旋转的机械能。用燃料或流水的能量带动原动机（称为涡轮机）就产生了机械能。这种机械能转换成电能是通过电动机来实现的，电动机直接连接在涡轮机轴上，由涡轮机驱动。因此，就在发电机的出线端获得电能，然后输送到需要它作功的地区。

发电需要的装置或机械（即原动机+发电机）统称动力设备。安置所有机械和其他辅助设施的建筑称为发电厂。

## 1 火电和水电

如上所述，涡轮机叶片是由燃料或流水的能量带动的。用燃料产生蒸气驱动蒸气涡轮机时，所产生的电称为火电。用于产生蒸气的燃料是一般燃料如煤、燃料油等，或是原子能燃料即核燃料。直接燃烧煤产生水蒸气，煤是最简便、最古老的一种燃料。柴油等也可以作为产生蒸气的燃料。原子燃料如铀、钍也可用于产生蒸气。用传统燃料如煤、燃料油等（称为矿物燃料）产生蒸气来带动水轮机时，这种发电厂一般称为普遍火力发电厂或热电厂。但当原子燃料用于产生蒸气时，这种发电站（基本上属于火力发电厂）称为原子能发电站或核电站。一般火力发电站是用锅炉产生蒸气的，而原子能发电站是用核反应堆和蒸气发生器代替锅炉产生蒸气的。这两种情况产生的电能称为火电。该系统称为火力发电系统。

然而，用流水的能量驱动水轮机时，所产生的电称为水电。这种系统称为水力发电系统，而发电厂称为水力发电厂或水电站。在水电系统中必须使具有一定势能和一定数量的水流流经水轮机。势能使水流动，驱动水轮机的叶片，这样与水轮机连接的发电机就发出电能。本章只涉及水力发电系统的内容。

## 2 水力发电站的分类

根据水力特性把水力发电站分为下列几种：①径流式电站；②蓄水式电站；③抽水蓄能电站；④潮汐电站。各类电站分述如下：

（1）径流式电站。这类电站是在河流上游无适宜的水库的情况下利用河流最小流量的电站。有时修建拦河堰坝，把水位提高并保持在预定的数值，只允许在很小的范围内变化。它可以单独为电站服务，或者主要为其他目标服务，兼顾电站。这种方案基本上是一种低水头方案，它仅适用于枯水季流量值得开发的常年性河流。

径流式电站通常具有很小的蓄水库容，有径流时方能利用。这个很小的蓄水库容是为满足每小时负荷的变化而设立的。当河道的来水流量大于发电需要时（在非峰荷期间），多余的水量就暂时蓄存在拦河建筑物上游的小水库中，以供峰荷期间使用。

径流式电站有诸多例子：楠加尔·海得尔运河的冈古瓦尔和科拉水电站，恒河的穆罕默德·普尔和帕持里水电站以及萨尔达运河的萨尔达水电站。

在灌溉渠道的跌水处修建的电站也属于径流式电站。

（2）蓄水式电站。蓄水式电站基本都有一足够大的上游蓄水库，贮存季风季节到干旱夏季的径流量，从而提供一个比枯季最小流量大得多的稳定流量。在这种设计方案中，水坝拦河修筑，电站可以布置在脚下，如巴克拉、希陶库德、里亨德工程等。电站也可能位于大坝下游很远的地方。在这种情况下，电站位于水库输水隧道的末端。输水隧道借助于压力水管与电站的机械装置连接，压力水管可能在地下（如迈吞和高勒工程），也可能在地上（如孔达工程）。

当电站位于大坝附近时，它一般采用低水头发电装置，这种电站称为集中落差式水力发电工程；但是当水流从大坝经过渠道、隧道或压力水管长距离输送到电站时，则称为分散落差式水力发电工程。

（3）抽水蓄能电站。抽水蓄能电站在峰荷期间发电，但在非峰荷期间，又把水从尾水池抽回到蓄水前池供以后使用。抽水机是由该系统其他电站的辅助电力驱动的。因而，这类抽水蓄能电站主要用于协助现有的火电站或别的水电站。

在峰荷期间，水从水库流入水轮机而产生电能。在非峰荷期间，利用其他电站的剩余电能，从尾水池抽水到前池，因而这个较小的电站为另一个较大的电站补充电能。在这样的系统中，同样的水量被一次又一次地重复利用，而没有被浪费。

为了利用在 15～90 米之间变化的水头，已制造出一种可逆式的水泵—水轮机，它既可作为水轮机也可作为水泵。这种可逆式水轮机可高效率地运转，有助于减少这类电站的投资。同样，同一种电力设备既可用作发电机，又可通过电极的互换而用作马达。这个系统中的设备非常有助于提高电力系统的负载系数。

（4）潮汐电站。用潮汐电站发电是近现代的成就。它是根据海水在高潮期上升、在落潮期下降的原理工作的。海水一日涨落两次。每次落潮周期大约是 12 小时 25 分。潮汐电站就是利用水位涨落的效益，换言之，就是利用高低潮之间的水位差进行发电的。为此，要修建一个水池，用隔墙和大海隔开，关在隔墙的孔洞里安装水轮机，就可以发电。

在高潮期间海水流入水池，驱动水轮机发电。在落潮期间，水又从水池流回海洋。只要安装一种在两个水流方向都能发电的特种水轮机组，就能利用流回海洋的水流进行发电。这类电站在潮差大的地方是很有用的。法国的朗斯电站就是这类电站的一个例子。那里的潮差达到 11 米。该站拥有九台机组，装机容量为 38000 千瓦。

根据水轮机的工作水头，可把水电站（或水电系统）分为下列几种：①低水头系统（落差小于 15 米）；②中水头系统（落差变化在 15～60 米之间）；③高水头系统（落差大于 60 米）。现分述如下：

（1）低水头系统。低水头系统使用的水头小于 15 米左右。径流式电站基本上属于低水头电站。在该系统中，修建拦河坝提高水位，电站或建在拦河坝的一端或建在坝的下游，离拦河坝有一定距离的地方，通过引水渠把水送往电站。

（2）中水头系统。中水头系统使用的水头变化在 15 米到 60 米左右。因此该系统基本上是一种大坝水库系统，尽管大坝的高度并不很大。在低水头和高水头系统之间，该系统在某些地方是有其优点的。

（3）高水头系统。高水头系统使用的水头大于 60 米。为了在上游蓄水和全年都能用水，要求建造有足够高度的大坝。已经发展的高水头系统的坝高已达 1800 米，该系统常见的例子如印度旁遮省的巴克拉大坝，印度北方邦的里亨德大坝，美国的胡佛大坝等。

高度较大的天然落差也可用来发电。这类动力开发的一般例子如印度的乔喀瀑布和美国的尼拉瀑布。

# 第二十五课 水 准 测 量

## 1 定义

水准测量是确定相互离开一段距离的各点高程的一种方法。通常由直接测得高差来确定高程。水准测量需要沿整个线路设置一系列的仪器站，在每一站上都要对已知高程点上的水准尺读取后视读数，并对未知高程点上的水准尺读取前视读数。

水准点（B.M.）是埋设在近乎永久性物体上的固定点，其位置和高程均为已知。水准点用来作为一定地区内的水准测量参考点，其高程由水准测量测定。全美国永久性水准点已由美国地质勘测局和美国海岸及大地测量局测设，是用铜板制成，埋设在石头或混凝土上，并标出高出平均海平面的高程。其他机构也设立了自用的永久性水准点。为了适应任何测量单位或施工单位的需要，设立了当地的水准点。这些水准点以假定的基准面为基准，采用天然或人工的物体，例如岩石、木桩、钉在树上或路面上的长钉以及漆在或刻在马路路沿石上的标记。

在某些地区，由于诸如地震、滑坡、地下水位下降、从油田中泵取石油、开矿或施工等原因而引起的地面升降，可能改变水准点的高程。

转点（T.P.）是在两个水准点之间的中间点，对这些点上的水准尺读取后视读数和前视读数。转点的特征要记入记录簿，但其位置则无需记录，除非准备再用此转点。水准点也可用做为转点。

后视读数（B.S.）是对已知高程点上的水准尺读取的读数。后视读数有时也称为正视读数。

前视读数（F.S.）是对待定高程点上的水准尺读取的读数。前视读数有时也称为负视读数。

仪器高度（H.I.）是仪器调平后仪器望远镜视线的高度。

在用经纬仪测量时，后视读数、前视读数和仪器高度等名词的含义与这里规定的不同。

## 2 实施步骤

在图 25.1 中，$B.M._1$ 为已知高程点（水准点），$B.M._2$ 为将在离 $B.M._1$ 一段距离处设立的另一水准点。现在需要测定 $B.M._2$ 的高程。立水准尺于 $B.M._1$ 上，安置水准仪于某一合适的地点，例如 $L_1$，点 $L_1$ 不要求设在连接 $B.M._1$ 和 $B.M._2$ 的直线上，只要在其附近即可。对 $B.M._1$ 上的水准尺读数取后视读数。然后司尺员前行，并按照观测员的指挥，在望远镜的视线范围内，在 $B.M._1$ 到 $B.M._2$ 的线路附近，选择某一合适的地点做为转点 $T.P._1$。希望（但并非必需如此）每一前视距离，如 $L_1$—$T.P._1$，大致等于相应的后视距离，如 $B.M._1$—$L_1$。立水准尺于转点上，并读取前视读数。观测员然后将仪器安置在另一合适地点，如 $L_2$，并对转点上的水准尺读取后视读数。此后，司尺员又前行，选择第二个转点 $T.P._2$；观测过程就如此重复进行，直到最后对终点 $B.M._2$ 上的水准尺读取前视读数。

从图 25.1 可知，后视读数加上后视点的高程，即得仪器视线的高程，从仪器视线高程减去前视读数，即得前视点的高程。同样，从某一点上的水准尺读取的后视读数与从下一点上的水准尺读取的前视读数之差，就等于这两点之间的高差。由此可见，所有后视读数的总和与所有前视读数的总和之差，就等于两个水准点之间的高差。

有时在进行隧道水准测量或房屋水准测量中，需要对一些高于仪器高度（H.I.）的测点上的水准尺读取读数。在这种情况下，要将水准尺倒立，并且在野外记录簿上在各后视读数前加上负号，在各前视读数前加上正号。

当要沿一条给定线路设置若干个水准点时，各中间水准点用来做为水准测量线路中的转点。常用延长水准测量线路并测回到初始水准点上的方法来校核各水准点的高程。终点回到初始点上的水准测量路称为闭合水准线路。如果线路的高差闭合差在容许范围之内，即可认为所有转点的高程的误差均在容许范围之内，但是这种闭合线路不能用来检核从闭合线路引伸出去的旁测点。

### 3　使后视距离和前视距离相等

在第 4-4 节中曾指出，如果前视距离等于相应的后视距离，则由于地球曲率和大气折光的影响（在相同的条件下）引起的任何读数误差都可消除。

在一般水准测量中，不需要设法使每一前视距离与相应的后视距离相等。各水准点之间的这些距离是否需要大致相等取决于所要求的精度。地球曲率和大气折光的影响并不大，除非在后视距离与前视距离之间存在着异常的差异。就使这些距离相等而论，仪器误差的影响似乎重要得多。视线和水准管轴线不绝对平等是很可能的。由于仪器校正的残余误差而产生的水准尺读数误差与从仪器到水准尺之间的距离成正比，并且对于后视读数和前视读数来说，符号都相同。由于后视读数要相加，而前视读数要减去，因此假如使各水准点之间的前视距离的总和与后视距离的总和相等，则仪器的误差可以消除。

在普通精度水准测量中，如使用的是经过仔细校正的仪器，后视距离与前视距离无需丈量，也无需设法，甚至无需通过估量，使二者相等。一般在长距离线路中，它们会趋于平衡。然而，像在高差大的两点之间进行水准测量时可能出现的那样，一系列很长的后视距离和相应有很短的前视距离，或者相反，会产生相当大的系统性误差。

在较高精度水准测量中，需要使水准点之间的后视距离和前视距离相等。在精度不太高的水准测量中，通常用步测来测定距离；在高山精度水准测量中，常用视距尺或倾斜测定仪来测定距离。

在进行上坡或下坡水准测量中，采用之字形线路，可在最少测站情况下使前视距离与后视距离相等。

使前视距离的总和等于后视距离的总和并不能完全消除地球曲率和大气折光的影响，相反，还必须使每一测站的前视距离等于相应的后视距离。

插图译文：图 25.1　水准测量。

## 第二十六课　施 工 和 设 备

许多大的承包商，特别在美国西部，都创业于为美国垦务局建造坝和运河。这些工程都按单价招标，而各项工程量系由业主（雇用的）工程师估算后列出。合同则按所有投标单价计算得出的最低标发包。

直到20世纪20年代中期，垦务局仍然保持有自己的施工力量，并喜欢用“计工”方式自己施工。垦务局内设有一个相当大而高效率的施工部门，负责用“计工”方式执行施工任务。这个部门由若干个直接向总工程师汇报的杰出的“施工总监”领导。他们根据局方预拟的计划和施工规范进行工作，并受局工程师的检查。他们申请设备及经总工程师批准雇用领班、熟练技工、机械工和工人，并经营自己的伙房和工房。造价都经过精心计算，工程质量也达到较高标准。

后来政策有了改变，即从用计工方式建造坝和发包小型结构物，转变到只要承包商的报价等于或低于计工方式的底价，所有工程都招标发包；这时，计工方式的施工组织才逐渐消失。

一些坝的营造单位完全出于需要，首次将工程的所有主要项目的施工程序和施工时间，用画图和列表的方式来安排。材料和设备则按照预先规划的进度表供应。从20世纪20年代末起，这一工作都靠辛勤的手工作业完成，直到有了计算机，并在1958年采用了“关键路线法”才有所改变。关键路线法需要将一个工程分解为一此分项活动，可以用网络图的形式来表示一项和另一项活动之间的顺序关系。

例如，伊太普工程施工作业的关键路线是通过下列几项主要作业，依次或同时进行的：①取得水泥供应的保证；②签订合同和确定适当的进点期限；③建造临时工房和运输道路；④开挖导流明渠；⑤修筑围堰；⑥供应和安装浇筑混凝土的栈桥及砂石料厂；⑦将坝和厂房的混凝土浇筑到139米高程，并安装进水口闸门；⑧购置并安装水轮机和发电机。

1930年前，施工设备全是小型的。施工工期都很长，承包商常缺少资金。因此，20世纪20年代后期产生了“合（资经）营企业”。新的、更大的和产量更多的设备不断地研制出来。例如，直到20世纪60年代绝大多数通用的石料装载机是动力铲。但从那时起，前端装料机的研制速度加快了。到20世纪70年代，前端装料机已在大多数需用场合取代了动力铲。但目前，又研制出15立方码的电铲，并正在伊太普工地上使用。

对于各种坝型来说，所需要的设备很不相同。设备的选择将取决于各个工程的具体情况。对于土石坝，常用的主要设备包括开挖、装料、运输和碾压机械，诸如履带式风钻、固定式压缩空气装置和输气系统、正向铲、反向铲、索铲、铲运机（单个或几个串列）、履带式拖拉机、底卸式拖车、后卸式卡车（越野的）、推土机、平土机、羊脚碾、轮胎碾、振动碾等。

对于混凝土坝，用于开挖和装料的许多设备，在开挖工作完成后，都能作为旧货出售。然而必须设置混凝土系统，包括骨料备制、堆料厂、混凝土配料和拌和设施、混凝土运送和浇筑设备。皮带运输机具有连续作业的优点，主要用于骨料处理和运输流态混凝土到浇筑地点。浇筑混凝土的最可靠和广泛使用的设备是：缆索起重机或有轨旋转式门机、混凝土吊罐和振动器等。

对于施工设备的讨论，如果不注意制造厂代表或销售工程师的作用，就不够全面。近来，制造厂代表或销售工程师通常是有学位的工程师，专门熟悉某一家制造厂所生产的设备能力。将这些专家仅仅作为售货员或“兜售商贩”来看待，是错误的。

## 第二十七课　水利规划中的工程经济

水利规划时要在所有许多可行性方案中考虑许多选择。一般来说，对这些方案的每一次选择都应根据经济原因进行。在选择进行之前，每一个需要认真考虑的方案都应当用货币单位表示出来。实际上，如果这些方案不能用货币单位表示，那么这些涉及诸多选择的项目就不能进行比较。例如，货币单位是各种性质不同项目诸如钢管、电的度数、技术和非技术劳动时数以及洪水造成的危害的减少等（方面）的惟一衡量单位。

工程经济方面的作者经常引用 Gen.John J.Garty 所提出的一些经典问题：究竟为什么做这件事？现在为什么做这件事？为什么用这种方式做这件事？这些问题被认为是一般问题的不同方面，它有意义吗？这些问题是 Gen.John J.Garty 在担任纽约电话公司总工程师时提出的。

根据这些问题建议，大多数工程计划都包括所有方案的主要部分，同时这些方案的许多次要方面又包含在每一个主要方案之中。例如，考虑用水电还是用热电来为一个电力系统增加需要的容量的问题。对水电方案，有许多不同的可能位置是可以利用的。在一个特定的地点，对一个引水坝可能有许多不同的可能设计；将水从引水点输送到发电厂，输水管有不同的类型和尺寸；在发电厂，水轮机的数量、类型和尺寸都存在不同的可能性，等等。每个次要方案可能会有它自己的附属方案。在分析的每个阶段，为了在所有可能的方案中做出合理的选择，考虑相对经济是必要的，这样才能制定出最优的总体规划。

当涉及以下步骤时进行经济研究方面的思考是有利的：

（1）在实物阶段，应对每一个可能有希望的方案进行鉴定并给予清楚的解释。

（2）具体到可实行阶段，对每一方案的实物评价应该转换成货币评价。一般而言，货币评价应由收入和支出组成，收入和支出将受到方案选择的影响。如同收入和支出的大小，评价也应包括日期。如果每一方案需要对建筑物和其他固定资产进行评估的话，就要求对其可变资本和残值进行评价。它也要求对研究期的长度和将要进行的经济分析的时间年限做出一个决定。

（3）通常，货币评价必须做为比较的依据。货币评价需要利用复利数学进行适当换算，这些转换应当利用最小回收利率做为利率。在特定的条件下，这样做是合适的。

（4）必须在所有的方案中做出一个选择（或为一个选择做出建议）。这一选择完全受以货币单位进行比较和其他对减少货币值行不通事件的影响。

一个年费用比较的简单例子列举如下。

对一条渠道的某个渠段，有两个可供选择的规划。规划 A 用隧洞，规划 B 用一段衬砌渠道和一段钢质渡槽。规划 A 中，隧洞初始投资估价为 450000 美元，年维护费估值为 4000 美元，隧洞估计使用寿命是 100 年。规划 B 各部分估计的初始投资和使用寿命为：渠道（不包括衬砌）120000 美元，使用寿命 100 年；渠道衬砌部分 50000 美元，使用寿命 20 年；钢渡槽 90000 美元，使用寿命 50 年；年维护费 10500 美元。经济分析中使用的利率为年利率 6%，研究期限为 100 年，所有残值假定忽略不计。两个规划之间没有估计收益差别，也没有其他预期的费用差别。（例如，两个方案之间没有期望的水量损失差别）

两个规划等效年度费用的比较如下：

规划 A：

隧洞资金偿还费：450000×0.06018=27081 美元

年维护费：4000 美元

总费用：31081 美元

规划 B：

渠道资金偿还费：120000×0.06018=7222 美元

渠道衬砌资金偿还费：50000×0.08718=4359 美元

渡槽资金偿还费：90000×0.06344=5710 美元

年维护费：10500 美元

总费用：27791 美元

前述表中，隧洞、渠道、衬砌和渡槽的初始费用值所被乘的复利因子取决于它们各自的估计寿命和 6%利率。将投资转换为等效年度费用的合适因子被称为资金偿还因子，该值可以用式 $i(1+i)^n[(1+i)^n-1]$ 进行计算，式中 $i$ 代表每年的利率（用小数表示），$N$ 表示估计的使用寿命。

当任何资金现值总量乘以利率为 $i$ 使用寿命为 $N$ 年的资本回收因子，乘积是足以准确偿还利率为 $i$ 使用寿命为 $N$ 年的现值总量的年度偿还费。例如，表中规划 B 中所示的 4359 美元是 50000 美元的资金偿还费，包括每年 6%的利息补偿。

规划 A 的 31081 美元的年费用与规划 B 的 27791 美元的年费用比较，可以作为工程经济分析中无数年费用比较的代表。通常，不同的设计要求不同的固定资产投资。在这个例子中，总投资额分别是 450000 美元和 260000 美元。额外的 190000 美元投资带来的优点是，它具有货币价值，如比较低的固定资产维护费和比较长的固定资产使用寿命。在这个例子中，年费用比较告诉我们，这些优点不足以证明这些额外投资是正确的。

所以我们应该选择规划 B，除非规划 A 有某些其他优点。

怎样为经济分析选择利率？

显然，计算费用时所采用的利率对年费用比较的结果有很大的影响。经济分析中利率的选择是为了指导工程项目的设计，利率的选择对设计能否被选用将会产生相当大的影响。用很低的利率，许多建议的投资（如规划 A 的隧洞）（表现出）好象是经济的，甚至是相同的投资，采用 6%或 8%的利率似乎也是非常浪费的。实际上，决定采用一个特定的利率，如规划 A 和规划 B 比较时使用 6%，这个被选定的利率是最小回收利率。

对私人企业，经济分析中使用的利率通常不应当小于一个值，该值不但能反映企业投资于新企业的资本总价值而且也能反映借入资本的总价值。

美国公共工程经济分析通常的作法已经达到采用的利率等于为上述公共工程借款的裸值。这个利率随着时间的不同和公共工程的不同而不同。1978 年联邦流域工程所采用的利率是 5.38%。

怎样估计水工建筑物的寿命？

国家税务局的公报发布了成千上万不同类型工业固定资产估计的平均寿命。水利工程某些部分的给定寿命（以年计）见表 27.1。

这样全国范围平均寿命的评价或许是有益的，即使他们不是在任何给定的情况下必须使用的最恰当的数值。况且，经济分析中采用的保守的估计寿命通常要比完全期望的使用寿命短。许多因素综合引起经济寿命比完全使用寿命短。一个有竞争力的私营产业有可能用 10 年作为有希望使用 20 年或更长时间的固定资产的偿还期。公共流域工程的造价通常以 100 年的寿命作为计算的依据。

对长寿命的固定资产，寿命估计的较大差异对年费用的影响比中等程度的利率差异所产生的影响小。例如，假定给定的寿命估计从 35 年增加到 100 年，同时所采用的利率从 5%增加到 6%；由于采用较高的利率而使年费用的增加值要比因使用较长寿命的估计值使费用减少的数值大很多。

表 27.1 水利工程各组成部分的使用寿命 单位：年

| | | | |
|---|---|---|---|
| 驳船 | 12 | 管： | |
| 木排拦河埂 | 15 | 铸铁管 | |
| 渠和沟 | 75 | 2～4 英寸 | 50 |
| 凝结沉淀池 | 50 | 4～6 英寸 | 65 |
| 施工设备 | 5 | 8～10 英寸 | 75 |
| 坝： | | 12 英寸以上 | 100 |
| 木笼 | 25 | 混凝土管 | 20 |
| 土坝、混凝土坝、圬工坝 | 150 | 钢管： | |
| 松散岩石坝 | 60 | 4 英寸以下 | 30 |
| 钢坝 | 40 | 4 英寸以上 | 40 |
| 过滤器 | 50 | 6 英寸石棉水泥管 | 50 |
| 渡槽： | | 输电线路 | 30 |
| 混凝土或圬工渡槽 | 75 | 拖船 | 12 |
| 钢渡槽 | 50 | 木狭板： | |
| 木渡槽 | 25 | 14 英寸以上 | 33 |
| 火力发电站 | 28 | 3～12 英寸 | 20 |
| 发电机： | | 抽水机 | 18～25 |
| 3000 千瓦以上 | 28 | 水库 | 75 |
| 1000～3000 千瓦 | 25 | 圆筒形水塔（竖管） | 50 |
| 500～1000 千瓦 | 17～25 | 水箱： | |
| 500 千瓦以下 | 14～17 | 混凝土制 | 50 |
| 给水栓 | 50 | 钢制 | 40 |
| 海运施工设备 | 12 | 木制 | 30 |
| 水表 | 30 | 隧洞 | 100 |
| 核电站 | 20 | 水轮机 | 35 |
| 压力管道 | 50 | 机井 | 40～50 |

# 第二十八课　土壤侵蚀与土壤保持

土木工程师都知道土壤运动的两种主要形式，即：土壤蠕动或土壤的缓慢流动；和称为滑坡、滑移或崩坍的快速流动。快速运动一般移动的土层较深，而土壤蠕动所移动的土层较浅。

但是比这两种形式更为重要的运动（至少就每年所移动的土壤数量而言）是土壤侵蚀和冲刷。土壤侵蚀是指水流从平常干燥的地面流过而冲走土壤；冲刷则指从溪水或河流的河床或者河岸把土壤或岩石冲走。正在生长中的树木下面的森林土壤一般不会被冲刷，因为树叶和枯木断枝像吸墨纸一样阻滞雨水并防止雨水侵袭土壤。枯叶抑制地面径流的主要作用是使雨水进入土壤的排泄渠道保持畅通。在没有受到这种方式保护的裸露地面，降雨时土壤的孔隙由于细小土粒流入很快就被堵塞。而水从地面流过而不进入土中，于是形成许多小沟，小沟愈冲愈深，终于成为真正能起破坏作用的深沟，这种冲刷生成的深沟称为沟壑。

土壤侵蚀对人类的第一个害处是侵蚀过的土地贫脊异常，因而不能生长庄稼，农民倾家荡产，不得不迁移到较好的土地上去。在美国，成千上万的农民都曾遭遇过这种情况。美国在过去100年中遭受土壤侵蚀的危害大概比任何其他国家都更为严重。土壤侵蚀对人类的第二个害处是由于水冲走土壤的速度有增无减，所以土壤侵蚀造成的水灾不仅更为严重而且更为频繁。在森林地区，降雨量的95%被吸收，而裸露的土壤则大约有95%的雨水流掉。几乎所有的土木工程建筑物、屋顶、道路和任何混凝土工程，其径流系数均为95%左右。

人们经常计算水灾期间土壤沉积在城市街道上所造成的损害，但是同样有相同的土壤从农田流失所引起的损害却没有人计算过，而这种损失对农民来说是同样严重的。

由于砍伐森林而引起的地表水快速流动，使原来很小的沟很快形成深沟，从而使沟中土壤侵蚀的速度增加并形成沟壑，防止土壤侵蚀的主要方法在于良好的耕作方法，诸如梯田，带状播种和不沿斜坡的上下坡方向而是水平地犁地（沿等高线犁地）等。在本书中我们不可能研究这些问题，但是土木工程师必须认识土壤侵蚀这个问题并力所能及地帮助农民。几乎所有的土木工程都使径流量增大，但是土木工程师可以通过使工程地面坡度平缓和植草栽种灌木或树从来努力减少土壤侵蚀，特别是在坡度陡峭土壤侵蚀可能最为严重的山区。

当在倾斜的地面或者斜坡的根部地面上修筑道路、铁路或者建筑物的基础时，每一位工程师都应注意地面滑坡的危险，并将其结构物设计得能够避免产生滑坡，但他也必须设法保持土壤。

美国土壤保持总署已经在林区养殖水獭来减小暴雨时河水的流量，因为这种动物建造堰坝横断水流，这种堰坝阻挡水势，使人类付出的费用很少，或者根本不需付任何费用。

在美国，主要公路保养工作中近1/3的工作量是针对土壤侵蚀的。在降雨量适中乃至降雨量大的所有丘陵地区的道路，也可采用类似的比例（或者更高的比例）。设计和建造涵洞时需要特别注意，因为水流不仅能冲刷涵洞上游的边缘，也能冲刷涵洞下游的边缘。假如涵洞坡度陡峭地斜向下游，可在离开涵洞处建造一个跌水式出口（图28.1）来防止冲刷。这种出水口是让水像瀑布那样流进一个不会被跌落的水流冲刷的混凝土结构物中。在涵洞上游建造跌水式进水口也是可能的。种植树木必须这样安排，勿使雪堆成很高的坡度。

小河堤岸的坡脚可以种植喜水树木一来保护它不受冲刷，这种方法费用低廉。树木削弱水的冲力并促使泥沙沉淀。过于裸露不能种植树木的河岸，可在上面铺放以灌木或灯心草制成的柴排加以保护，泥沙可以积聚在柴排里，而植物就得以生长。为了紧急保护朝向激流的裸露的河岸，可以铺砌大的块石，也可以铺砌预制混凝土板或者袋装干拌混凝土。

堤岸的防波堤保护稍有不同，短的防波堤或突堤是建在河里面的，堤岸的上游正冲蚀着它。防波堤的缺点是即使它们确实能防止它们所在的堤岸被水冲刷，但它们却可能促使它们对面的堤岸遭受相应的冲刷。防波堤可用木桩打入河床，或将大量石头沿河岸倾斜方向倒入河里，甚至可以用垃圾如旧汽车车身。

河流过的下游区域是宽阔平缓的洪积平原，这里可能要求土木工程师像建防洪工程，河堤就是这样的防洪建筑物，它通常是由一条长的廉价的土堤构成，土堤尽可能建在距河岸比较远的地方。这样就允许洪水分布在河流和堤防之间的土地上。河流与堤防的距离越大，防洪堤内承纳的洪水就越多，防洪堤就越有用。这样被淹没的土地通常都是牧场或其他农田，他们没有被洪水严重损坏。

# 第二十九课 地下水与气候变化

许多事实证明，温室排放的气体使得大气浓度增加。我们目前正处于一个气候变化的时期。自从19世纪后期以来，全球平均温度上升了0.3～0.6℃，全球海平面上升10～25cm（政府间气候变化组织IPCC，1995）。据IPCC报道，预计下世纪全球温度上升幅度要比过去10000年实测上升温度大。由于温度的上升，水文循环也会发生较大的变化，降水量和蒸发量也会发生相应的变化。恶劣的气象事件，洪水爆发、以及干旱事件都有可能发生。

在加拿大，尤其是加拿大哥伦比亚（BC），关于气候变化对水文循环的影响已经做了许多研究，研究内容大多数是预测气候变化对地表水的影响，重点是冰川径流和河道流量相互关系的研究。关键性参数变化（例如，降水和径流）对于地下水反应灵敏度的研究开展的相对较少，尽管在加拿大地下水占淡水供水的相当大的部分。国际文献中有关气候变化（未来变化趋势）对地下水资源影响的研究报道也不多。

本项研究之目的是灵敏度评价，识别加拿大Grand Forks地区气候变化（变化趋势）对非承压含水层地下水的影响。Grand Forks地区位于哥伦比亚的中南部，Vancouver以东522km，沿加拿大—美国边境线，见图29.1。

该地区属于半干旱地区，地下水是当地居民生活和灌溉用水的主要水源，为了制定长期水利管理决策，识别气候变化对地下水的潜在影响是十分必要的。

饮用水占哥伦比亚地区地下水供水的22%，地下水也是本省许多农业区的灌溉用水。因此，研究气候变化对地下水系统的影响是十分重要的。地下水是水文循环的一部分，它受补给变化的影响，也受到降水量和蒸散发变化的影响，也会受地下水和地表水系统相互联系特性的影响，以及灌溉用水的影响。

关于地下水补给问题，已经开展了大量的研究。Vaccaro（1992）用逐日补给模型估计美国华盛顿洲哥伦比亚高原半干旱地区Ellensburg流域的补给变化。他根据三种地下水计算模型的平均值，使用了随机气象生成仪对未来趋势进行预测，还研究了土地开发利用前（草原、山艾和森林）后（灌溉和非灌溉土地）变化的影响。在土地未开发利用条件下，地下水补给量增加10%，原因是降水量的增加和春季积雪融化入渗量的增加。对于现状土地利用条件下，由于灌溉入渗的增加，地下水补给量比土地开发利用前高出4倍。在气候变化条件下，尽管增加了降雨量，现状土地利用下的地下水补给量将减少40%。原因是灌溉水量蒸发增加和积雪入渗量减少，但就其影响范围来说，现状土地利用条件下要比土地开发利用前的影响范围小。

BC地区内，在晚春和早夏期间，由于积雪融化，大面积地下水得到补给。因此补给量取决于冬季降水量、春季融雪量以及补给历时。如果补给历时不变，降水量的增加会增加补给量，虽然大部分降水量不会成为地下水，但是也会直接汇入河流。地下水储量受土壤空隙度的影响，过多的水量不会全部下渗到地下水，而是作为地表径流流走。类似地，入渗强度受地下岩土材料类型的影响，大的暴雨也不会全部下渗。地下水补给量的变化对地下水位也有一定的影响，相应地，也会对河流的基流产生影响。

在BC地区的许多地方，高富水含水层位于冲积峡谷，该含水层由松散的冰川或河流沉积物组成。许多含水层的地表有河流通过。由于地表沉积物的高透水性，地下水—地表水的相互联系密切。因此，河流流量大小和时间的变化，与气候变化有关，也会对地下水位产生相当大的影响。

# 第三十课 灌溉运行评价

灌溉运行评价是提高农业用水效率、缓解缺水矛盾的必要条件。从 1996 ～2000 年，我们对位于西班牙南部 Andalusia 的 Genil–Cabra 灌区（GCIS）运行进行了系统性评价。该灌区灌溉面积 7000 公顷，分为 843 个单元，种植作物种类繁多，以向日葵、棉花、大蒜、橄榄树为主。灌区灌溉由承压系统供水，手动喷灌是该灌区最流行的灌溉方法。本项研究以单元用水资料和模拟模型为基础，分别用 6 个运行指标来评价 GCIS 灌区的灌溉用水和管理经济运行状态。模拟模型模拟了不同地块的水量平衡过程，计算了优化灌溉方案，并与实际灌溉方案对比分析。在运行指标中，平均灌溉供需比（实测灌溉供水与模拟优化需水之比）呈年际变化，从 0.45～0.64，说明该地区为缺水灌溉。如果考虑降雨因素，一年的供需比增加到 0.87，但是极端干旱年份为 0.72，说明干旱年份没有得到充足的灌溉水量补充。

由于其他因素（如城市用水、旅游、娱乐和环境需水）需水量的增加，未来灌溉可利用水量有可能减少。西班牙的淡水资源需水量估计为 $35\times10^5$ 立方米/年，其中有 70%用于灌溉。其余用水占 30%。另外，政府部门预计在未来 10 年内，西班牙南部的灌溉需水量将增加 17%。

进一步提高西班牙灌区的水利管理和现代化以及工程改造是实现高效用水的重要目标。西班牙只有 27%的灌区（总面积约 915000 公顷）建成年限少于 20 年，有 37%的灌区年限超过了 90 年。近年来，水利管理组织重视了灌区系统现代化建设和灌区工程改造，但是对于提高灌溉管理重视不够。

改善灌区水利管理需要以灌溉运行评价为起点。以水文模型为基础，计算机模拟是灌溉运行评价的有用工具。灌溉农业水文循环模拟模型很多，有些是经验性的，有些是函数型的，也有的是过程描述型的。另外，为了便于资料收集和空间分析，近年来开发了一些工具如遥感、地理信息系统，这些工具与水文模型相结合，可以满足灌溉计划效果评价的需要。

许多作者提出了不同的评价指标，用来反映灌溉系统运行特征、评价运行效果、提出改进灌溉效率和水分生产率的建议。也可以用这些指标对系统能力进行定量分析，为实现灌区目标提供依据，或者以灌溉系统潜力为标准，用这些指标评价灌溉系统现状运行情况。

不同类型的运行指标有：①水量平衡；②经济、环境和社会目标；③系统维护。许多作者将这些指标用于：①评价运行趋势；②不同灌溉计划方案的对比分析；③资源优化；④确定一个灌区的整体平衡与局部高效之间的协调方案。用来计算水量平衡运行指标的模型有一维模型，物理模型，水文模型以及非常简化的模型，如 FAO 提出的方法。当编制灌溉计划时，计算运行指标的输入信息通常是从总供水资料和由作物种植面积计算用水资料而得到的。这些信息往往带有很大的不确定性，不能用来进行总体计划以下的深度分析。所以，需要针对不同目的灌溉计划分层评价，当缺乏计划层以下的资料时，这是惟一的可行性方法。

本书目的是利用田间用水资料和模拟模型，对灌区的灌溉运行开展系统性评价。所选灌区位于西班牙南部 Andalusia 的 Genil–Cabra 灌溉系统。选择该灌区的原因是在四次灌溉时期内，它具有精确的用水资料和各个单元的作物布局资料。

# 第三十一课 招标、开标和授予合同

招标文件应该清楚地说明，根据所提供的货物或者工程建筑的性质，合同是以分项价格（完成的建筑或者提供的货物），还是以总价格为基础授予的。

单个合同的规模和范围取决于项目的大小、性质和地点。对于包含种类繁多的工程建筑和设备的项目，如电力、供水，或者工业项目，通常是就土木工程和不同的工厂厂房、设备等主要项目的供应和安装分别签订合同。

另一方面，如果一个项目需要有一些相似的、独立的土木工程或设备，合同的方式应该多样化，供投标人选择，这样不仅能吸引较大的公司，也能吸引较小的公司。应该允许各大小承包商或制造商，对单个合同或一组类似的合同随意进行投标。所有的单标或组标应该同时开标和进行估价，以决定哪个单标或组标对借款人最合算。

在招标之前，要把所需工程建筑或货物的详细设计包括技术要求和其他招标文件准备就绪。然而，对于包括规划、设计和管理的整套施工合同或者大的综合性工业项目的合同，事先准备好的技术规范也许是不合乎需要的。这时，就需要采用一种人为两步的程序，先是进行不确定费用技术招标，以便于进行技术性说明及安排，随后再提出带价格的标书。

投标所需要的准备时间应该视合同的规模和复杂程度而定。一般来说，国际投标从招标之日起应有不少于 45 天的时间。如果涉及大型土木工程，一般来说，从招标之日起，应该给予不少于 90 天的时间，以使有可能中标的投标人在投标前能到施工现场进行调查。而到底需要给多少时间，应该视工程项目的具体情况而定。

投标人投标截止的日期、时间和地点，以及开标的日期、时间和地点都要在招标时公布，所有的投标都要在规定的时间打开。在截止时间以后收到的投标，都要原封不动地退回。除非迟到的原因不在于投标人，而且在截止期后接受其标书并不使他比其他投标占便宜。投标应该在公开场合打开，投标人的名字，每个投标的总金额，包括要求他们或者允许提出的可供选择的投标，都要在开标时大声朗读，记录在案。

一般来说，不应该延长投标的有效期限。如果在特殊情况下，需要延长投标期限，就要在终止日期之前向所有投标人提出要求。投标人有权拒绝延长期限，并不得因此而失掉其投标保证金。对那些同意延长投标期限的投标人，即不要求，也不允许修改他们的投标。

在公开开标以后，宣布合同中标者之前，不允许把有关对投标进行审查、说明和评价以及关于中档的推荐意见等情况，告诉投标人或者与这些程序没有正式关系的人。开标之后，任何投标人都不允许修改投标。只接受不改动投标实际内容的说明。借款人可以要求投标人对其投标加以说明，但不能要求投标人改动实际内容或报价。

开标以后，就应该查明投标中有无计算错误，投标是不是基本上回答了招标文件提出的要求，有没有提供所需要的保证，文件是否都签了字，投标文件在其他方面是否完整。如果投标基本上没有回答招标文件提出的要求，或者包含不能允许的保留，就要加以拒绝，除非这是根据招标文件的要求，或者文件所允许的供选择的参考投标。随后，就要进行技术分析，对每个投标进行评价，并进行比较。

借款人或者其顾问要准备一个对投标的评价和比较的详细报告，在里面要写明其所依据的具体理由。并应写明将合同授予谁，或者所有投标都不能接受。

要在规定的投标有效期内，把合同授予以最低价投标的，而且他在能力和财力方面符合相应标准的投标人。通常，不应当要求投标人承担技术要求中没有规定的责任或者工作，或者要求其修改投标，并以此作为授予合同的条件。

# 第三十二课 如何撰写科学论文

## 1 题目

作者在准备论文题目时，应该记住一个明显的事实；论文的题目将被成千上万人读到。能完整地读完整篇论文的人，如果有的话也只是少数几个人。大多数读者或者通过原始期刊，或者通过二次文献（文摘或索引）来阅读论文的题目。因此，题目中的每一个词都应该仔细地推敲，词与词之间的关系也应该细心处理。

论文的题目是一个标记，它不是一个句子。因此，它不必像通常的句子那样具有主语，动词和宾语结构。实际上，它比句子简单（或者，至少通常比较短），但是词的排列顺序则更为重要。

题目中每个词的含意和词序，对于可能阅读期刊目录中论文题目的读者来说是很重要的。这对于所有可能使用文献的人，包括通过二次文献查找论文的人（可能是大多数）也同样重要。因此，题目不仅仅作为论文的标记，它还应该适合于化学文摘，工程索引，科学引文索引等机器索引系统。大多数索引和摘要都采用“关键词”分类法。因此，在确定论文题目时，最重要的是作者应该提供能够正确表达文章内容的“关键词”，也就是说文章的题目用词应该限于既容易理解，又便于检索，又能使文章的重要内容突出的那些词。

## 2 摘要

摘要应该是论文的缩写版本。它应该是论文各主要章节的简要总结。一篇写得好的摘要能使读者迅速而又准确地了解论文的基本内容，以决定他们是否对此论文感兴趣，进而决定他们是否要阅读全文。摘要一般不超过 250 个单词，并应该清楚地反映论文的内容。许多人将阅读原始期刊上或工程索引、科学引文索引或者另一种其他二次出版刊物上刊登的摘要。

摘要应该阐述该项研究工作的主要目的和范围；①描述所使用的方法；②总结研究成果；③阐述主要结论。结论的重要性可以由其通常被三次提出来而表明：在摘要中提一次，在引言中提一次，在讨论中再提一次（可能会更详细些）。

摘要决不应该提及论文中没有涉及的内容或结论。在摘要中不要引用与该论文有关的参考文献（在极少的情况下除外，例如对以前发表过的方法的改进）。

## 3 引言

我们已经结束了对正文前面的内容的讨论，现在就应该开始研究如何写正文了。我应该指出，虽然题目和摘要放在论文的前面，但是一些有经验的作者往往是在写完论文后再写题目和摘要。然而，在写论文时，你心里必须要有（如果没有写在纸上的话）一个暂定的论文的提纲。你还应该考虑读者的水平，并据此确定哪些术语和方法需要定义或描述，和哪些不需要定义或描述。

当然，正文的第一部分应该是引言。引言的目的是向读者提供足够的背景知识，使读者不需要阅读过去已经发表的有关此课题的论文，能够了解和评价目前的研究成果。在引言中也应该提出该项研究工作的理论基础。最重要的是，你应该简要地说明写这篇论文的目的。应该慎重地选择参考文献以提供最重要的背景资料。

一篇好的引言，应该遵守下列原则：①首先应该尽可能清楚地提出所研究的问题的性质和范围；②为了适应读者的需要，应该对有关文献进行评述；③应该阐述研究方法；如果认为有必要的话，则应该阐述选择特定方法的理由；④应该阐述研究的主要成果；⑤应该阐述由结果得出的主要结论。不要让读者有悬念，应该让他们知道证据的由来。

## 4 材料与方法

在论文的第一节即“引言”中，你叙述了研究工作中所使用的方法。如果需要的话，不可能对你所选择的某一种特定方法的理由加以说明。

在“材料与方法”一节中，你应该提供详细的试验细节。“材料与方法”一节的主要目的是描述试验过程和提供足够的细节，以使有能力的研究人员可以重复这个试验。许多（可能是大部分）读者可能会略过这一节不看，因为他们在引言中已经知道了你所使用的一般方法，也可能他们对试验细节不感兴趣。但是，认真撰写这一节是非常重要的，因为科学方法的核心就是要求你的研究成果不仅有科学价值，而且也必须是能够重复的；为了判断研究成果能否重复，就必须为其他人提供进行重复实验的依据。不太可能重复的实验是不可取的，必须具有产生同样或相似结果的潜力，否则你的论文的科学价值就不大。

当你的论文受到同行们审核时，一个好的审稿人会认真地阅读“材料与方法”这一节。如果他确实怀疑你的实验能够被重复，不管你的研究成果多么令人敬畏，这个审稿人都会建议退回你的稿件。

如上所述，在描述研究方法时，你应该提供足够的细节，以便使有能力的研究人员能够重复这些实验。如果你的方法是新的（未发表过的），那你就应该提供所需要的全部实验细节。然而，如果一个实验方法已在正规的期刊上发表过，那么只要给出参考文献就可以了。

## 5 结果

现在我们进入论文的核心部分——数据。论文的这部分被称为“结果”。

“结果”一节通常由两部分组成。首先，你应该对实验作全面的叙述，提出一个“大的轮廓”，但不要重复已经在“材料与方法”一节中提到的实验细节。其次，你应该提供数据。

当然，如何提供数据并不是那么容易的事。很少有可以直接将实验笔记本上的数据抄到稿件上的事。最重要的是，在稿件中你应该提供有代表性的数据，而不是那些无限重复的数据。

“结果”一节应该写得清晰和简练，因为“结果”是由你提供给世界的新知识组成。论文的前几部分（“引言”、“材料与方法”）告诉人们的是你为什么和怎样得到这些结果的；而论文后面的部分（“讨论”）则告诉人们这些结果意味着什么。因此，很明显，整篇论文都是以“结果”为基础的。所以“结果”必须以确切而清晰的形式给出。

## 6 讨论

与其他章节相比，“讨论”一节所写的内容更难于确定。因此，它是最难写的一节。不知你是否知道，尽管许多论文中的数据正确，而且能够引起人们的兴趣，但是由于讨论部分写得不好也会遭到期刊编辑的拒绝。甚至更为可能的是，如果在“讨论”中所作的阐述使得数据的真正含义模糊不清，变得难懂，则会使论文遭到退稿。

一个好的“讨论”章节的主要特征是什么呢？我认为它应该包括以下几个方面：

（1）设法给出“结果”一节中的原理，相互关系，和归纳性解释。应该记住，一个好的“讨论”应该对“结果”进行讨论和论述，而不是扼要重述；

（2）要指出任何的例外情况或相互关系中有问题的地方，并应该明确提出尚未解决的问题。决不要冒着很大的风险去采取另一方式，即试图对不适合的数据进行掩盖或捏造；

（3）要说明的解释你的结果与以前发表过的研究结果有什么相符（或者不相符）的地方；

（4）要大胆地论述你的研究工作的理论意义以及任何可能的实际应用；

（5）要尽可能清晰地叙述你的结论；

（6）对每一结论要简要叙述其论据。

在描述所观察的事物之间的相互关系时，你并不需要得出一个广泛的结论。你很少有能力去解释全部真理；更常见的是，你尽最大努力所做的就是像探照灯那样照耀在真理的某一方面。因为你在这个方面的真理是靠你的数据来支持的；如果你将你的数据外推到更大的范围，那就会显得荒唐，这时甚至连你的数据所支持的结论也可能会受到怀疑。

当你叙述你的这一点点真理的意义时，要尽可能地简单。最简单的语言可以表达最多的学识；啰嗦的语言或者华丽的技术术语常常只表达肤浅的思想。

# 参 考 文 献

1 金芷生等．工业与民用建筑专业英语．南京：江苏科学技术出版社，1985

2 Changjiang Water Resources Commission. The Three Gorges of the Yangtze River—A Guide for Tourists. China Water Power Press，1997

3 Ren Decun. A General Introduction of the Yellow River——to Learn the Yellow River From Here. The Yellow River Publishing House，1999

4 王秉钧，郭正行．科技英汉汉英翻译技巧．天津：天津大学出版社，1999

5 亢树森．科技英语写作与翻译教程．西安：陕西科学技术出版社，1998

6 李立功，朱宝贵等．水利电力英语．南京：河海大学出版社，1989

7 史中庸，龙腾锐．土木建筑系列英语——给水与排水．北京：中国建筑工业出版社，1987

8 宽本．水利水电科技英语阅读和翻译．第二版．北京：水利电力出版社，1985

9 贾艳敏，吕景山．土木工程专业英语．哈尔滨工业大学出版社，1998

10 Daniel Hillel.Advances in Irrigation.Academic Press，1982